MICROWAVE-ENHANCED POLYMER
CHEMISTRY AND TECHNOLOGY

MICROWAVE-ENHANCED POLYMER CHEMISTRY AND TECHNOLOGY

Dariusz Bogdał and Aleksander Prociak

Dariusz Bogdał, PhD, Professor, Institute of Polymer Science and Technology, Cracow University of Technology in Kracow, Poland.

Aleksander Prociak, PhD, Asst. Professor, Institute of Polymer Science and Technology, Cracow University of Technology in Kracow, Poland.

Blackwell Publishing Professional
2121 State Avenue, Ames, Iowa 50014, USA

Orders: 1-800-862-6657
Office: 1-515-292-0140
Fax: 1-515-292-3348
Web site: www.blackwellprofessional.com

Blackwell Publishing Ltd
9600 Garsington Road, Oxford OX4 2DQ, UK
Tel.: +44 (0)1865 776868

Blackwell Publishing Asia
550 Swanston Street, Carlton, Victoria 3053, Australia
Tel.: +61 (0)3 8359 1011

First edition, 2007

Library of Congress Cataloging-in-Publication Data

Bogdał, Dariusz.
    Microwave-enhanced polymer chemistry and technology / Dariusz (Darek) Bogdał, Aleksander Prociak. — 1st ed.
      p. cm.
    Includes index.
    ISBN-13: 978-0-8138-2537-3 (alk. paper)
    ISBN-10: 0-8138-2537-7 (alk. paper)
    1. Polymers. 2. Polymerization. 3. Microwaves—Environmental aspects.
4. Green technology. I. Prociak, Aleksander. II. Title

    QD281.P6B6 2007
    668.9 dc22
                                                          2006100548

The last digit is the print number: 9 8 7 6 5 4 3 2 1

DEDICATION

This book is dedicated to my wife Małgorzata; and to our sons Mikołaj and Grzegorz; and to their grandmother and my mother Antonina.
  —Dariusz Bogdał

This book is dedicated to my wife Ewa and to our children Tomasz and Agnieszka.
  —Aleksander Prociak

TABLE OF CONTENTS

PREFACE

The emergence of more efficient and environmentally benign methods of performing chemical synthesis has created microwave-enhanced chemistry, one of the most significant events of the 1990s with tremendous potential for the 21st century. Recently, it was realized that such features of processing under microwave irradiation—short processing time, energy efficiency, high yield, low waste, and use of alternative solvents or solvent-free conditions—could play an important role in the development of "green chemistry" methods in the future.

Polymer chemistry and technology, aside from ceramics processing, forms what is probably the largest single discipline in microwave technology, and the methods and procedures used are certainly among the most developed. For example, a remarkable achievement of the late 1960s was the application of microwave irradiation for the continuous vulcanization of extruded rubber and the discontinuous vulcanization of molded rubber articles. The microwave-assisted vulcanization of rubber compounds used industrially since the 1970s is the most important application of microwave heating to polymeric materials in terms of number of installed plants.

Although a great deal of papers have been published on the processing and curing of polymers and polymeric materials under microwave irradiation, and the application of microwaves to polymer synthesis is still growing. There exists only a limited number of full resources on the microwave-assisted polymer synthesis and processing; therefore, this subject was taken as the main objective of the book.

This book is intended for polymer chemists and engineers in both industry and academia to relate novel approaches of polymer chemistry and processing under microwave irradiation and thus spread ideas of microwave technologies. It is hoped that the book will also serve as an introduction to the field for industrial chemists without prior training in microwave-assisted synthesis and processing of polymers. The book covers background and scientific data, discussions of processes and product properties in comparison with existing technology, and the status of current research. The book should also prove interesting to chemistry students whose knowledge of the subject is more advanced.

The first chapter provides basic knowledge about the interaction of materials with microwaves and explains the dielectric heating mechanism and, therefore, why some materials can be effectively heated under microwave irradiation. Also, a short description of equipment (i.e., generators, waveguides, and applicators) required for microwave processing is presented.

The second chapter deals with an overview of polymerization processes under microwave conditions. The intention of this chapter is to emphasize differences and limitations and to provide examples of the most spectacular applications of microwave irradiation in comparison to conventional processes.

In the third and fourth chapters, the synthesis of thermoplastic polymers and thermosetting resins is described, respectively. Examples include most syntheses in addition to microwave systems that are used by different research groups, and the scale of the experiments will give readers a deeper feeling about the microwave-assisted processes.

The fifth chapter describes interactions during the bonding of polymers, joining of thermoplastics (welding), thick composites, and higher performance polymeric composite reinforced with carbon and glass fibers under microwave irradiation.

The sixth and seventh chapters present the application of microwave irradiation for modification of natural polymers (polysaccharides) and synthesis of raw materials based on vegetable oils for thermosetting resins, together with degradation processes of plastic waste and chemical recycling of polyurethanes and polyesters.

The eighth chapter discusses aspects of commercialisation. Microwaves do not provide a universal solution to all problems but should be considered whenever all other processes fail to solve an industrial problem, in which case the advantages of microwaves become unique and offer considerable savings compared to other existing processes.

During the past four decades, the overriding need for close cooperation between the user and the manufacturer of microwave equipment has become apparent. Thus, a real cross-disciplinary approach has to be considered to fully understand all the limitations and advantages of microwave processing. Improper application of microwave irradiation will usually lead to disappointment, whereas proper understanding and use of microwave power can bring even greater benefits than expected.

Dariusz Bogdał and Aleksander Prociak
Kraków, 2006

MICROWAVE-ENHANCED POLYMER CHEMISTRY AND TECHNOLOGY

1

FUNDAMENTALS OF MICROWAVES

Microwave frequencies occupy the electromagnetic spectrum between radio frequencies and infrared radiation with the frequencies of 300 GHz to 300 MHz, which corresponds to the wavelengths of 1 mm to 1 m, respectively (Fig. 1.1). Their major applications fall into two categories, depending on whether they are used for transmission of information (telecommunication) or transmission of energy. However, the extensive application of microwaves in the field of telecommunication (e.g., most of the wavelengths in the range of 1 cm to 25 cm are used for mobile phones, radar, and radar line transmissions) has caused only specially assigned frequencies to be allocated for energy transmission (i.e., for industrial, scientific, or medical applications). Currently, to minimize interferences with telecommunication devices, these household and industrial microwave applicators are operated only at a few precise frequencies with narrow tolerance that are allocated under international regulations. For example, the most common microwave applicators (i.e., domestic microwave ovens) use the frequency of 2.45 GHz. This is probably why most commercially available microwave reactors devoted for chemical use operate at the same frequency; however, some other frequencies are also available for heating (Thuery, 1992).

Interaction of Microwaves with Materials

When a piece material is exposed to microwave irradiation, microwaves can be reflected from its surface if it is an electrical conductor (e.g., metals, graphite, etc.), can penetrate the material without absorption in the case of good insulators with good dielectric properties (e.g., quartz glass, porcelain, ceramics), and can be absorbed by the material if it is a lossy dielectric (i.e., a material that exhibits so-called dielectric losses, which in turn results in heat generation in a quickly oscillating electromagnetic field, such as water) (Fig. 1.2).

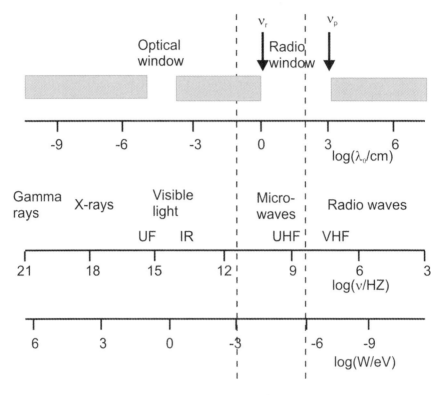

Fig. 1.1 Spectrum of electromagnetic radiation: λ_0, wavelength in free space; W, hν quantum energy; ν_r, lowest resonance frequency in the rotational spectrum of water; and ν_p, plasma frequency of the ionosphere. Reprinted from Kaatze, U. 1995. Radiat. Phys. Chem. 45:539, with permission.

When a strongly conducting material (e.g., a metal) is exposed to microwave radiation, microwaves are largely reflected from its surface (Fig. 1.2a). However, the material is not effectively heated by microwaves; in response to the electric field of microwave radiation, electrons move freely on the surface of the material, and the flow of electrons can heat the material through a resistive (ohmic) heating mechanism (Fig. 1.3a). In the case of insulators (e.g., porcelain), microwaves can penetrate the material without any absorption, losses, or heat generation. They are transparent to microwaves (Fig. 1.2b). For some dielectrics, the reorientation of either permanent or induced dipoles during passage of microwave radiation, which is electromagnetic in nature, can give rise to absorption of microwave energy and heat generation due to the so-called dielectric heating mechanism (Fig. 1.2c). Dependent on the frequency, the dipole may move in time to

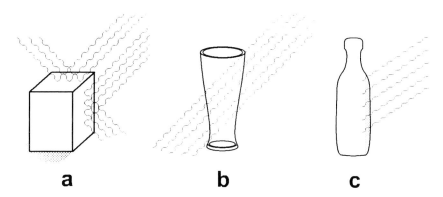

Fig. 1.2 Interaction of microwaves with different materials: (**a**) electrical conductor, (**b**) insulator, and (**c**) lossy dielectric.

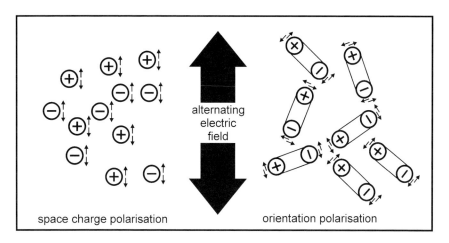

space charge polarisation   orientation polarisation

Fig. 1.3 Interaction of charge particles and dipolar molecules with electromagnetic radiation: space charge polarization (**a**) and orientation-polarization (**b**), respectively.

the field, lag behind it, or remain apparently unaffected. When the dipole lags behind the field (polarization losses), interactions between the dipole and the field lead to an energy loss by heating (i.e., by dielectric heating mechanism), the extent of which is dependent on the phase difference of these fields (Fig. 1.3b).

In fact, the electric field component of microwave radiation is responsible for dielectric heating mechanisms because it can cause molecular motion by either migration of ionic species (conduction mechanism) (Fig. 1.3a) or rotation of dipolar species (dipolar polarization mechanism) (Fig 1.3b). In a microwave field, the electric field component oscillates very quickly (at 2.45 GHz, the field oscillates 4.9×10^9/sec), and the strong agitation, provided by cyclic reorientation of molecules, can result in an intense internal heating that can lead to heating rates in excess of 10 °C/sec when microwave radiation of a kilowatt-capacity source is used (Metaxas et al., 1983).

In practice, most good dielectric materials are solid and examples include ceramics, mica, glass, plastics, and the oxides of various metals, but some liquids and gases can serve as good dielectric materials as well. For example, deionized water is a fairly good dielectric; however, possessing polar molecules (i.e., a dipole moment) can couple efficiently with microwaves to lead to heat generation due to polarization losses. Thus, such substances that are counted among dielectrics but exhibit some polarization losses that result in the dielectric heating are also called dielectric lossy materials or in general lossy materials. On the other hand, *n*-hexane, having a symmetrical molecule, does not possess a dipole moment and does not absorb microwaves.

To apply microwaves to carry out chemical processes, it is most important to find at least one component that is polarizable and whose dipoles can reorient (couple) rapidly in response to changing electric field of microwave radiation. Fortunately, a number of organic molecules and solvents fulfill these requirements and are the best candidates for microwave applications. The first step is to analyze the reaction components together with their dielectric properties, among which the most important is dielectric constant (ε_r), sometimes called electric permeability. Dielectric constant (ε_r) is defined as the ratio of the electric permeability of the material to the electric permeability of free space (i.e., vacuum), and its value can derived from a simplified capacitor model (Fig. 1.4).

When the material is introduced between two plates of a capacitor, the total charge (C_0) stored in the capacitor will change (C) (Eq. 1.1). The change depends on the ability of the material to resist the formation of an electric field within it and, finally, to get polarized under the electric field of the capacitor.

$$\varepsilon_r = \frac{C}{C_0} \tag{1.1}$$

where C_0 is the capacitance of the capacitor with vacuum and C is the capacitance of the capacitor with the material. Thus, dielectric constants (ε_r) that determine the charge holding ability of the materials are characteristic for each substance and its state and vary with temperature, voltage, and, finally, frequency of the electric field. Dielectric constants for some common materials are given in Table 1.1.

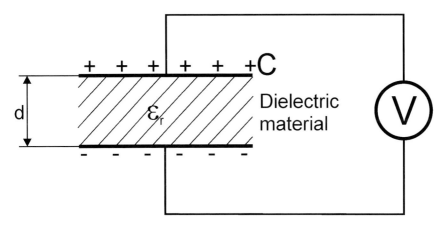

Fig. 1.4   An electrical capacitor consisting of two metal plates separated by an insulating material called a dielectric.

Air has nearly the same dielectric constant as vacuum (ε_r = 1.00059 and 1.00000, respectively). Polar organic solvents (i.e., water, acetonitrile, ethyl alcohol) are characterized by relatively high values of dielectric constants and, in turn, can be heated by dielectric heating mechanism under microwave irradiation. Nonpolar organic solvents (i.e., benzene, carbon tetrachloride, *n*-hexane) have low dielectric constants and, in fact, show negligible heating effects under microwave irradiation. Most plastics range in the low values of dielectric constants (i.e., between 2 and 3); therefore, some

Table 1.1.   Dielectric constants (ε_r) of some common materials at 20 °C.

Material	Dielectric constant (ε_r)	Material	Dielectric constant (ε_r)
Vacuum	1	Titanium dioxide	100
Air (1 atm)	1.00059	Water	80
Air (100 atm)	1.0548	Acetonitrile	38
Glass	5–10	Liquid ammonia ($-78°$)	5
Quartz glass	5	Ethyl alcohol	25
Porcelain	5–6	Benzene	2
Mica	3–6	Carbon tetrachloride	2
Rubber	2–4	Hexane	2
Nylon	3–22	Plexiglass	3
Paper	1–3	Polyvinyl chloride	3
Paraffin	2–5	Polyethylene	2
Soil (dry)	2.5–3	Teflon	2
Wood (dry)	1–3	Polystyrene (foam)	1.05

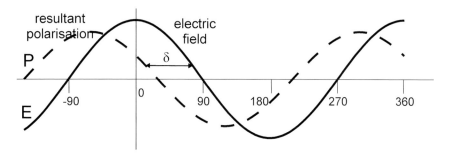

Fig. 1.5   Phase diagram—the lag of polarization (P) behind the electric field component (E) by the phase δ.

of these materials, besides glass and quartz glass, are used to manufacture reaction vessels for microwave application due to their good chemical as well as temperature resistance (e.g., Teflon, PEEK). Thus, heating a material in microwave ovens is based upon the ability of some liquids as well as solids to absorb and to transform electromagnetic energy into heat.

For an ideal dielectric between the plates of capacitor (Fig. 1.4), there is no lag between the orientation of the molecules and the variations of the alternating voltage. The displacement current is 90° (δ) out-of-phase with the oscillating electric field (Fig. 1.5).

The relevant phase diagram (Fig. 1.6a) shows that for a dielectric material where the molecules can keep pace with the field changes, no heating occurs. There is no component of the current in-phase with the electric field; that is, the product $E \times I$ is zero because of the 90° phase lag between the field and the current.

If the frequency of the electromagnetic radiation is pushed up into the microwave region ($\approx 10^9$ Hz), the rotations of the polar molecules in the material begin to lag behind the electric field oscillations. The resulting phase displacement δ (Fig. 1.6b) acquires a component $I \times \sin \delta$ in phase with δ the electric field, and so resistive heating occurs in the medium; this is described as dielectric loss and causes energy to be absorbed from the electric field. Because the dipoles are unable to follow the higher-frequency electric field oscillations, the permeability falls at the higher frequency and the substance behaves increasingly like a nonpolar material (Gabriel et al., 1998).

However, in the case of highly oscillating electric fields for which the loss angle (δ) differs significantly from 90°, the material possesses a dual role and act both as a dielectric and conductor. Because sinδ is an in-phase current component, dielectric constant is turned into complex permeability (ϵ^* = ϵ' − j ϵ''), which is a measure of the ability of dielectric materials to absorb and store electrical potential energy (Fig. 1.6c).

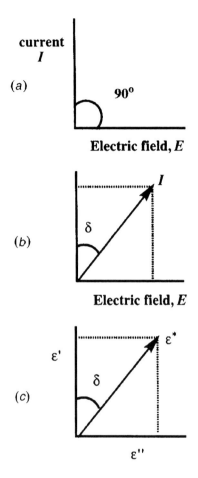

Fig. 1.6 Phase diagrams for (**a**) an ideal dielectric where the energy is transmitted without loss; (**b**) where there is a phase displacement δ and the current acquires a component $I \times \sin \delta$ in-phase with the voltage and consequently there is a dissipation of energy. In (**c**), the relationship between ε^*, ε', and ε'' is illustrated; $\tan \delta = \varepsilon''/\varepsilon'$. Reprinted from Gabriel, C., Gabriel, S., Grant, E.H., Halstead, B.S.J., Mingos, D.M.P. 1998. Chem. Soc. Rev. 27:213, with permission.

Like dielectric constant (ε_r), which is commonly used to describe good isolators, the real permeability (ε') component characterizes the ability of the material to be polarized by the electric field and thus the ability of microwaves to propagate into the material. At low frequencies, this value reaches its maximum (i.e., ε_r) because the maximum amount of energy can be stored in the material. The imaginary part of the complex electric permeability (ε'') is usually called the loss factor and indicates the ability of the

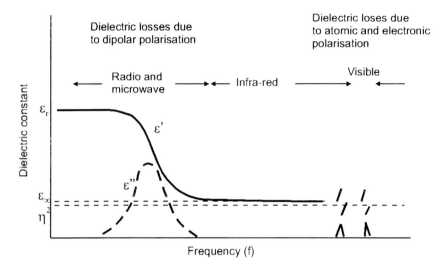

Fig. 1.7   Relative real permeability (ε') and loss factor (ε'') depending on angular frequency of electromagnetic waves. Reprinted from Metaxas, A.C., Meredith, R.J. 1983. Industrial Microwave Heating. Peter Perigrinus. London, with permission.

material to dissipate the energy (i.e., the efficiency of conversion of electromagnetic radiation into heat). The loss factor (ε'') reaches the maximum when the real permeability (ε') gradually decreases (Fig. 1.7). Obviously, it depends on the molecular structure of the material, on the frequency at which ε'' reaches its maximum, and on how distinctly this maximum is exhibited.

The values of real permeability (ε') and loss factors (ε'') for some common organic solvents are given in Table 1.2 (Mingos et al., 1991).

Systematic solvent studies have disclosed the existence of a linear relationship between the polarity of the molecules (evaluated by their dielectric constants) and the temperature increase induced by microwave irradiation. The effect of microwave activation is therefore strongly dependent on the solvent nature, rather weak when nonpolar solvents are employed, considerable with polar solvents, and large with strongly polar solvents, which can be at the origin of very rapid heating (Bram et al., 1992).

Another commonly used term to characterize materials under microwave conditions is the loss tangent, (tan δ), which is calculated as the ratio of ε'' and ε' (Eq. 1.2), and this parameter even better describes the ability of a material to absorb microwave energy.

$$\tan \delta = \frac{\varepsilon''}{\varepsilon'} \qquad (1.2)$$

Table 1.2. Real permittivities (ε') and loss factors (ε'') of organic solvents at different frequencies at 20°C.

Solvent	3×10^8 Hz ε'	ε''	3×10^9 Hz ε'	ε''	1×10^{10} Hz ε'	ε''
Water	77.5	1.2	76.7	12.0	55.0	29.7
0.1 M NaCl	76.0	59.0	75.5	18.1	54.0	30.0
Methanol	30.9	2.5	23.9	15.3	8.9	7.2
Ethanol	22.3	6.0	6.5	1.6	1.7	0.11
Propanol	16.0	6.7	3.7	2.5	2.3	0.20
Butanol	11.5	6.3	3.5	1.6	0.2	—
Hexane	1.97	—	1.97	2×10^{-4}	1.97	9×10^{-3}
Carbon tetrachloride	2.2	—	2.2	9×10^{-4}	2.2	3×10^{-3}

Reprinted from Mingos, D.M.P., Baghurst, D.R. 1991. Chem. Soc. Rev. 20:1, with permission.

The loss tangent and loss angle as well as real permeability (ε') and the loss factor (ε'') are always dependent on working frequency and temperature. Effects of frequency and temperature and the complex permeability are shown in Figure 1.8. According to the frequency of the electromagnetic field, three general changes with temperature of the dielectric properties can be observed: (1) real permeability (ε') and the loss factor (ε'') decrease with temperature, (2) real permeability (ε') and/or the loss factor (ε'') pass through a maximum, and (3) real permeability (ε') and the loss factor (ε'') increase with temperature (Stuerga et al., 2003). As in case 1, some materials (e.g., water) become a poorer microwave absorber with rising temperature, but other lossy materials can become a better microwave absorber with rising temperature like in case 2. The rate at which temperature of such lossy materials rises is proportional to the increase of the loss factor (ε''); i.e., "the hotter they get, the quicker they get hotter." The rapid rise in the loss factor (ε'') with temperature is therefore the major factor in thermal runaway and temperature nonuniformity.

The values of tan δ of some common organic solvents at room temperature together with boiling points of these solvents and their temperatures after 1 minute of microwave irradiation are presented in Table 1.3.

It can be seen (Tables 1.2 and 1.3) that for optimum coupling (i.e., effective heating), a balanced combination of moderate ε', to permit adequate penetration, and high loss ε'' as well as tan δ are required. The higher the loss tangent, the better is the transformation of microwave energy into heat. However, water absorbs microwaves efficiently at the frequency of 2.45 GHz and has a high ability to transform microwave energy into heat, but it

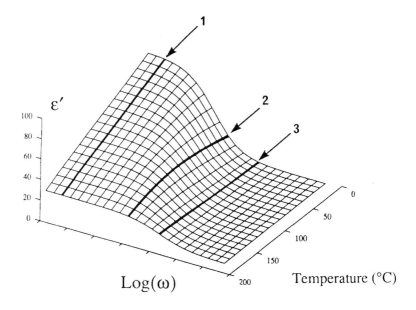

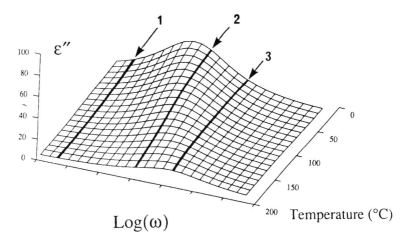

Fig. 1.8 Change of complex dielectric permeability (i.e., ε′ and ε″) with frequency and temperature. Reprinted from Stuerga, D., Delmotte, M. 2003. Wave-material interaction, microwave technology and equipment. In: Microwaves in Organic Synthesis. Loupy, A. (Ed.). Wiley-VCH. Weinheim, with permission.

Table 1.3. Loss tangent tan δ at 2.45 GHz (room temperature), boiling points, and temperatures of common organic solvents after about 1 minute of irradiation in a microwave cavity (50 mL of solvent, 1 minute, power 560 W).

Solvent	Solvent temperature (°C)	Boiling point (°C)	Loss tangent tan δ
Water	81	100	0.12
Methanol	65	65	0.66
Ethanol	78	78	0.94
Propanol	98	97	0.76
Butanol	106	117	0.57
Ethyl acetate	73	77	0.06
Acetone	56	56	0.05
Acetic acid	110	118	0.17
DMF	131	153	0.16
Methylene chloride	41	41	0.04
Dioxane	53	101	—
Hexane	25	68	0.02
Carbon tetrachloride	28	77	—

Reprinted from Bogdal, D. 2005. Microwave-Assisted Organic Synthesis, One Hundred Reaction Procedures. Elsevier. Amsterdam, with permission.

is worth stressing that alcohols are much better microwave absorbers with values of the loss tangent (tan δ) a couple of fold higher than the loss tangent (tan δ) of water. Therefore, owing to lower values of heat capacity than water (c.f., water and ethyl alcohol heat capacities are 4.18 and 2.31 J/g K, respectively), alcohols can be heated with much higher heating rates than water to reach their boiling points in a shorter time (Table 1.3). Both water and alcohols can be added in a small amount to reaction mixtures in order to improve their dielectric properties (i.e., increase absorption of microwaves). Nonpolar solvents (e.g., hexane, carbon tetrachloride) are not good subjects for microwave activation; however, small amounts of them are added to reaction mixtures, in particular under solvent-free systems, in order to improve their temperature homogeneity during microwave irradiation (Bogdal, 2005).

The loss tangent (tan δ) of some other common materials are presented in Table 1.4, while the loss tangent of water as the function of frequency is presented in Figure 1.9 together with ε' and ε''. It is apparent that appreciable values of the loss factor (ε'') exist over a wide frequency range. For example, for water, the most effective heating as measured by ε'' reaches its maximum at approximately 20 GHz, while most microwave ovens operate at a much lower frequency (i.e., 2.45 GHz). The practical reason for the

Table 1.4. Loss tangents (tan δ) of different materials at 25 °C (2.45 GHz).

Material	Loss tangent tan δ ×10^{-4}	Material	Loss tangent tan δ ×10^{-4}
Quartz	0.6	Plexiglass	57
Ceramic	5.5	Polyester	28
Porcelain	11	Polyethylene	31
Phosphate glass	46	Polystyrene	3.3
Borosilicate glass	10	Teflon	1.5

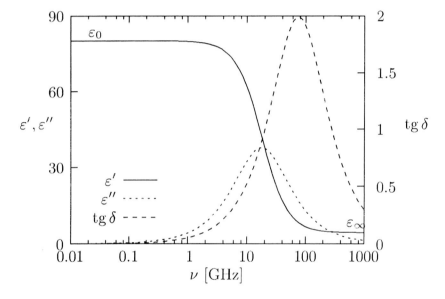

Fig. 1.9 The loss tangent of water (tan δ) as the function of frequency together with ε′ and ε″. Reprinted from Gabriel, C., Gabriel, S., Grant, E.H., Halstead, B.S.J., Mingos, D.M.P. 1998. Chem. Soc. Rev. 27:213, with permission.

lower frequency is to heat the material throughout its interior. This is so that the radiation is not totally absorbed by the first layer of the material that it encounters and may penetrate farther into the material, heating it more evenly. In other words, if the frequency is optimal for maximum heating rate, the microwaves are absorbed in the outer region of the material and penetrate only a short distance ("skin effect") (Metaxas et al., 1983).

If the electric field is assumed to be uniform throughout the volume, the amount of power (*P*) that is absorbed per unit volume is given by Equation

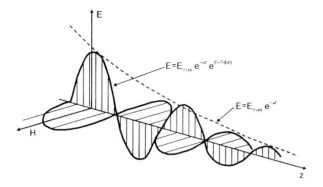

Fig.1.10 Propagation of electromagnetic waves in a lossy material. Reprinted from Metaxas, A.C., Meredith, R.J. 1983. Industrial Microwave Heating. Peter Perigrinus. London, with permission.

1.3, where ω is the angular frequency and E is the potential gradient of the electric field:

$$P = \frac{1}{2} P_{max} E_{max} \omega \sin(\delta)$$

While energy is absorbed within the material, the electric field decreases as a function of the distance from the material surface. Therefore, Equation 1.3 is valid for only very thin materials. Upon penetrating a lossy dielectric, the waves are attenuated and the power is dissipated after traversing for a certain distance (Fig. 1.10). Thus, for the processes that are carried out under microwave conditions on larger scales, another important factor that describes the heating process is the penetration depth (D_p).

The penetration depth is defined as the distance from the sample surface where the absorbed power is $1/e$ of the absorbed power at the surface. Beyond this depth, volumetric heating due to absorbing of microwave energy is negligible. The D_p is proportional to the wavelength of the radiation and depends on the dielectric properties of the material. For lossy dielectrics ($\varepsilon''/\varepsilon' < 1$), the D_p can be calculated from Equation 1.4.

$$D_p = \frac{\lambda_0}{2\pi} \frac{\sqrt{\varepsilon'}}{\varepsilon''} \tag{1.4}$$

where λ_0 is the length of electromagnetic wave.

Materials with relatively high values of the loss factor (ε'') are characterized by low values of the D_p, and, therefore, microwaves can be totally ab-

Table 1.5. Microwave (2.45 GHz) penetration depth (D_p) in some common materials. Reprinted from: Kubel (2005) Industrial Heating 43 with permission.

Material	Temperature (°C)	Penetration depth (D_p) (cm)
Water	25	1.4
Water	95	5.7
Ice	−12	1100
Paper, cardboard	25	20-60
Wood	25	8-350
Rubber	25	15-350
Glass	25	35
Porcelain	25	56
Polyvinylchloride	20	210
Epoxy resin (Araldite)	25	4100
Teflon	25	9200
Quartz glass	25	16,000

Reprinted from Kubel, E. 2005. Ind. Heating January: 43-53, with permission.

sorbed within the outer layers of these material. For example, the penetration depth of water is 1.4 and 5.7 cm at 25 and 95 °C, respectively (Table 1.5). This means that during the experiments in a water solution on larger scales, only some parts (outer layers) of the reaction mixture interact with microwaves to generate heat, which is then transported into the rest of the mixture conventionally (i.e., by heat convection and/or conduction mechanism). On the other hand, microwaves can penetrate and pass through the samples of the materials with low values of the loss factor (ε''). However, in the case of chemical processes, this should not be treated as a drawback because, due to the design of the microwave reactors, unabsorbed radiation passing through the sample is mostly reflected back and, finally, absorbed on later passes, which can provide enough energy to initiate a process (Bogdal, 2005).

The heating rate (i.e., temperature increase) of the material under microwave irradiation also depends on the shape and size of the sample. For instance, the study of the thermal behavior of alumina indicates that the temperature of the sample during microwave irradiation depends on the amount of the sample (Fig. 1.11). In a multimode cavity, the maximum temperature was obtained for approximately 200 g of alumina because it had a larger surface and, in turn, could absorb more of the microwave energy per weight unit than, for example, 100 g of alumina. In contrast, the 500-g sample of alumina was heated with lower heating rate than the 200-g sample, but in this case the most probable explanation is that 500 g of alumina

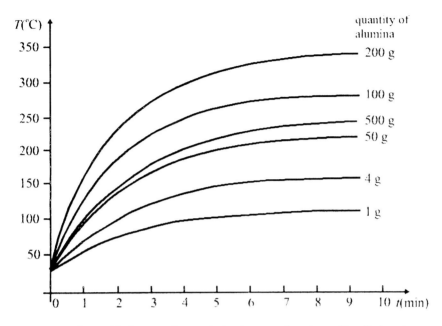

Fig. 1.11 Thermal behavior of neutral alumina as a function of irradiation time and quantity. Reprinted from Bram, G., Loupy, A., Majdoub, M., Gutierrez, E., Ruiz-Hitzky, E. 1990. Tetrahedron 46:5167, with permission.

consumed the entire microwave energy provided by the reactor, which in fact could be less per weight unit than for 200 g of alumina. Most likely, the amount of sample that would be heated most efficiently by microwaves under such experimental conditions was somewhere between 200 and 500 g of alumina. In turn, 4 g of alumina appeared as the minimum amount for an appreciable thermal effect to be observed, while 1-g sample of alumina was unable to reach 100 °C even after a relatively long irradiation time (≈20 minutes) (Bram et al., 1990).

Eventually, the sample size, penetration depth, and heating rate are strongly coupled during microwave irradiation and may finally result in more homogeneous or heterogeneous heating of the material, which in turn can result in overheating of the material and creation of so-called hot spots in the latter case (Bogdal et al., 2005).

It is worth stressing that microwaves in comparison with conventional heating methods are the means of volumetric heating of the material that gives rise to a very rapid energy transfer into the material being heated. In conventional heating, heat flow is initiated on the material's surface and the rate of heat flow into the center is dependent on the material's thermal properties and the temperature differentials. A conventional oven is re-

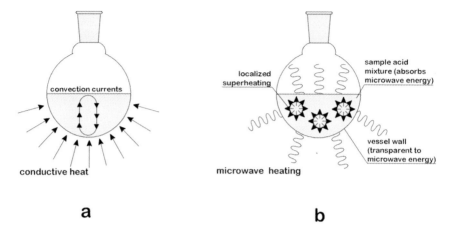

conductive heat

microwave heating

a

b

Fig. 1.12   Different heating mechanisms for conventional (a) and microwave heating (b).

quired to be heated to temperatures much higher than is required by the material itself (Fig. 1.12).

Microwave Equipment

Microwave devices that are dedicated to carry out chemical processes are similar to other microwave systems that consist of a microwave power source (generator), transmission line (waveguide) that delivers microwaves from the generator into an applicator, and microwave applicator (cavity) (Fig. 1.13). The microwave applicator is the means by which microwaves can be efficiently coupled with the material. One of the frequently used applicators is a multimode microwave cavity, the same as those used in microwave ovens. Other applicators in the form of horn antennae, radiators, and single-mode waveguide resonators can also be used to enable efficient coupling of the energy with the sample. A reliable measuring system is necessary to at least monitor temperature of the process, and for processes running at elevated pressures, appropriate pressure control is required, too.

Microwave Generators

The main types of microwave power sources are magnetrons and klystrons. Magnetrons, which are commonly used in microwave ovens, are mass produced and thus are cheap and easily available on the market. Therefore, it is common practice to use the same magnetrons for laboratory and industrial microwave processing. In general, magnetrons are vacuum devices

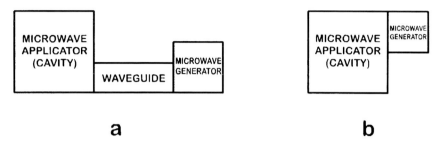

Fig. 1.13 Schemes of microwave devices with (a) and without (b) transmission line.

consisting of an anode and a cathode, and the anode is kept at a higher potential than the cathode. As soon as the cathode is heated, electrons are emitted from it and are accelerated toward the anode by the electric field. At the same time, external magnets mounted around the magnetron anode block create the magnetic field parallel to the axis of the cathode. This magnetic field forces the electrons to rotate around the cathode before they can reach the anode. The rotating electrons form a rotor moving around the cathode synchronously in the way that they decelerate and thus transform their energy into microwave oscillations in the cavities cut in the anode blocks. A typical anode block consists of an even number of small cavities that form a series of microwave circuits, which are tuned to oscillate at a specified frequency dependent on the dimensions and shape of the cavities. Finally, microwave energy from one of the resonant cavities is coupled to the output transmission line, usually terminated with an output antenna (Fig. 1.14).

Magnetrons belong to a wider class of microwave devices called traveling wave tubes (TWTs), with the difference that the magnetron has no input for microwaves and therefore can only be used as a generator, while a TWT can be used as an amplifier as well. This is because in a TWT an electron beam is moving along the so-called slow wave or traveling wave structure, which can retard propagation of electromagnetic waves so they can conveniently (from the point of view of energy exchange) interact with and gain energy from the electrons moving in the same direction. Consequently, there is another type of microwave amplifier called backward wave tubes (BWTs) in which the electrons and the wave travel in opposite directions. This can be more energy efficient for power sources in the submillimeter range of microwaves. An extensive description of microwave sources like TWTs, klystrons, disk-seal triodes, and power grid tubes can be found in a number of books on industrial microwave applications (Metaxas et al., 1983; Thuery, 1992).

20   *Microwave-Enhanced Polymer Chemistry and Technology*

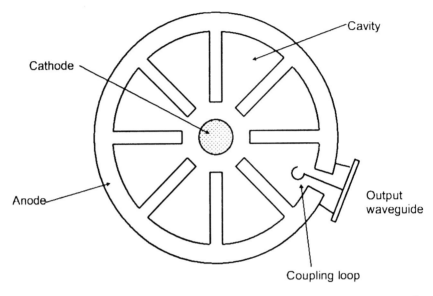

Fig. 1.14   Schematic diagram of a magnetron shown in cross section. Reprinted from National Academy Press. 1994. Microwave Processing of Materials. National Academy Press, with permission.

Transmission Lines (Waveguides)

Microwaves can be easily transmitted through various media. Hence, an applicator can be remote from the power source and connected to it via the transmission line, in which microwaves can be propagated using three types of modes. In transverse electromagnetic mode (TEM), all the components (i.e., magnetic and electric) are transverse, which is similar to, if not an approximation of, the radiation in free space. Both electric and magnetic components in the propagation direction (z) are missing. Microwaves in TEM mode can propagate between two parallel wires, two parallel plates, or in coaxial lines and are applied in general to low-power systems (≈ 10 kW) (Fig. 1.15) (Metaxas et al., 1983; Thuery, 1992).

The transverse electric (TE) and transverse magnetic (TM) are typical modes propagated inside metallic waveguides, which are basically hollow conducting pipes having either a rectangular or circular cross section. In the TE and TM modes, the component of the electric field and the magnetic field in the propagation direction is, respectively, missing. In general, every electromagnetic wave in rectangular or cylindrical waveguides can be described as a linear combination of the TE and TM modes, and, in turn, the indexes in the propagation modes (e.g., TE_{10}) describe particular eigenvalues that define wave variation in a given direction. For example,

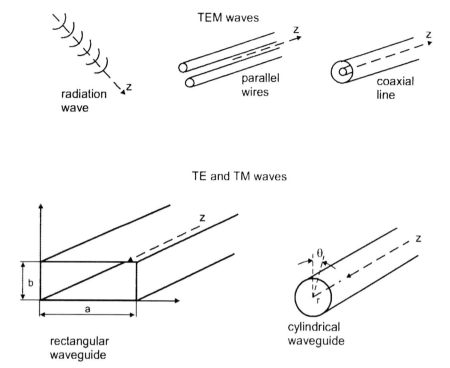

Fig. 1.15 Waveguides for propagating transverse electromagnetic (TEM), transverse magnetic (TM), and transverse electric (TE) waves. Reprinted from National Academy Press. 1994. Microwave Processing of Materials. National Academy Press, with permission.

a TE_{10} mode in rectangular waveguide specifies fundamental TE mode with one half-wavelength in the x-direction and zero half-wavelengths in the y-direction. If the waveguide section forms a cavity, then a third index describes how many half-wavelengths would fit along this section. For instance, TE_{103} is a cavity having one half-wavelength along x-direction, three half-wavelengths along the z-direction, and zero variation of electric field in y-direction (Fig. 1.16) (Metaxas et al., 1983).

Microwave Applicators (Cavities)

Microwave applicators may appear in many different shapes and dimensions, and in fact their design is critical to processes run under microwave conditions since within applicators the microwave energy must be efficiently coupled to the material. A number of the most commonly used microwave applicators belong to three main types:

22   *Microwave-Enhanced Polymer Chemistry and Technology*

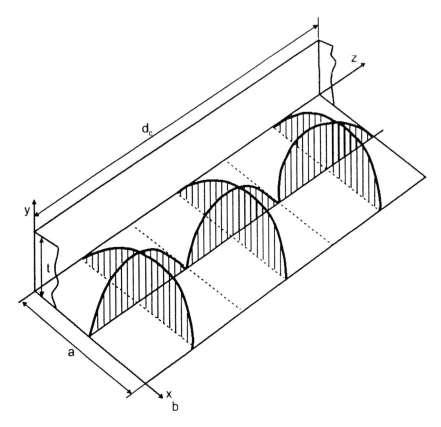

Fig. 1.16   Development of standing waves: TE_{103} resonant cavity electric field pattern. Reprinted from Metaxas, A.C., Meredith, R.J. 1983. Industrial Microwave Heating. Peter Perigrinus. London, with permission.

- Multimode cavity
- Single-mode cavity including TEM structures
- Traveling wave applicators

The important criterion characterizing applicators is the maximum electrical field strength generated within the material. It is worth stressing that it is possible to map the field strength distribution within a cavity by moving, for example, a small water load around it and recording the rate of temperature rise, but the results of such an exercise apply only to water. Once another target material is placed in the cavity, the field distribution is changed self-consistently.

The simplest applicator is a rectangular metal box that can accommodate the target material. While microwaves are applied into such a box via a

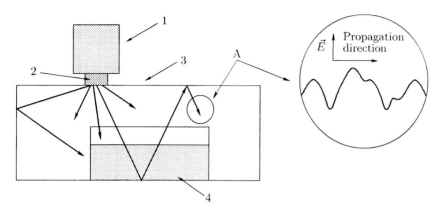

Fig. 1.17 Multimode microwave reactor: 1, magnetron; 2, rotating deflector; 3, multimode cavity; 4, reaction vessel; A, nonregular shape of electromagnetic waves as a superposition of a number of waves.

Table 1.6. Comparison of different microwave systems.

Parameters	Microwave R-220A (Sharp)	Synthwave 402 (Prolabo CEM)	Ethos MR Milestone
Microwave mode	Multimode	Single-mode	Multimode
Microwave power	800 W, pulsed	300 W, unpulsed	1000 W, pulsed/ unpulsed
Volume of cavity	15.7 dm^3	≤ 0.25 dm^3	42.8 dm^3
Power density	approx. 50 W/dm^3	≤ 1200 W/dm^3	approx. 23 W/dm^3
Reaction scale	≤ 100 g (dry reaction)	≤ 100 g	up to 3000 g (depending on reactor type)

Reprinted from Nuchter, M., Muller, U., Ondruschka, B., Tied, A., Lautenschlanger, W. 2003. Chem. Eng. Technol. 26:1207., with permission.

waveguide, the microwaves undergo multiple reflections from the walls. The reflected waves interfere and, in so doing, their superposition establishes a distribution of electrical field strengths within the internal space (including the material), which within the band of frequencies covered by a magnetron corresponds too many different possible modes of oscillations. For this rea son such a metal box (cavity) is called a multimode microwave applicator (cavity) (Fig. 1.17). The most common example of this device is a cavity in domestic microwave oven usually with maximum microwave power less than 1200 W and average power density in an empty cavity around 50 W/L (Table 1.6).

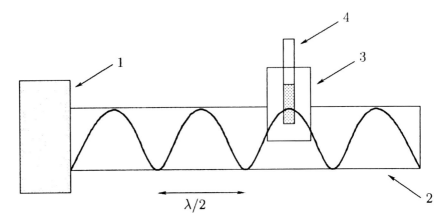

Fig. 1.18 Single-mode microwave reactor: 1, magnetron; 2, waveguide; 3, single-mode cavity; 4, reaction vessel.

The field distribution within the materials that is contained in a multimode applicator (cavity) depends not only on the real permeability (ε') and loss factor (ε'') of the material but also on the size and location of the sample within the applicator. In this respect, a multimode device is best suited to a very lossy material occupying a relatively large volume (more than 50%) of the applicator. For low and medium loss materials occupying less than about 20% of the applicator volume, the temperature rise in the material at best will be nonuniform, and at worst potentially damaging "hot spots" will occur as the result of extremely high local fields. However, by incorporating a mode stirrer (a rotating microwave deflector) and continuously rotating the material placed in the applicator, temperature uniformity within the material can readily be improved. It must be underlined that multimode systems have a self-tuning feature (i.e., they change field distribution and field strength dependent on the size, shape, and absorption ability of the load). Disadvantageous in the use of multimode applicators (cavities) for scientific studies is that because the spatial distribution of field strength is unknown, the possibility to generalize the results from a particular investigation is compromised, making it slightly difficult to attempt reliable reproducibility and as a consequence scaling-up the elaborated process. By far a very efficient applicator, particularly for the syntheses on a small scale ≤50 mL), is a single-mode resonant cavity, in which only one mode of microwave propagation is permitted and hence the field pattern is well defined so the material can be positioned accordingly (Fig. 1.18).

A single-mode cavity may be either cylindrical or rectangular, and the simplest (i.e., physically smallest) single-mode cavity operates in the TM_{010} mode, in which the electrical field strength is constant along the main axis

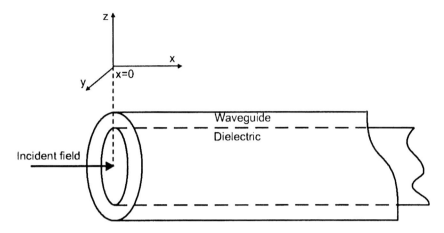

Fig. 1.19   Traveling wave applicator.

without variation of angle and is only a function of radius. The radius of the loaded cavity can be determined by solving the electric and magnetic field equations within the cavity under assumption of dimensions and particular boundary conditions. A rectangular single-mode cavity consists of a section of waveguide that is terminated with a noncontact plunger, which in turn can tune the effective cavity length in order to obtain resonant frequency and conditions. The mode of operation of such a device is, typically, TE_{10} with the target material positioned in a region of high field strength. The major limitation of this device is that the length of the load (material) must be less than one half-wavelength (about 6 cm for 2.45 GHz microwave frequency); otherwise, the wave standing along the material would release heat in a periodic manner leading to a periodic heating pattern along the material. However, this can be overcome by, for example, moving the plunger or the material in a reciprocating fashion so that the time-averaged field seen by the material is smoothed. Usually single-mode applicators (reactors) are characterized by maximum microwave power of about 300 W, but owing to much smaller size than multimode cavities, the average power density of single-mode cavities can reach 1200 W/L (Table 1.6).

One of the solutions for the problems mentioned with the application of single-mode and multimode applicators (cavities) is the use of a traveling wave applicator (Fig. 1.19). However, it is not a common solution for the construction of microwave reactors. A traveling wave applicator consists of a section of a slow-wave structure with traveling wave propagating along it and coupled to the material in a spatially distributed manner. The wave is traveling, enabling efficient coupling to a sample of the material having a bigger volume or larger surface than is passing through the structure,

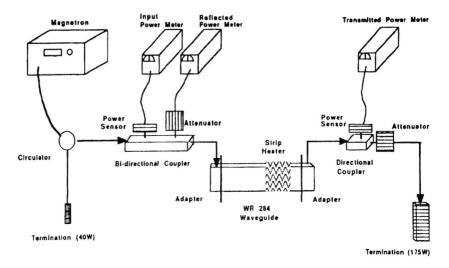

Fig. 1.20　Schematic of the experimental setup for the traveling wave applicator with stub tuner. Reprinted from Chen, M., Siochi E.J., Ward, T.C., McGrath, J.E. 1993. Polym. Eng. Sci. 33:1092, with permission.

from the other end of which is launched a microwave source. As microwaves pass along the slow-wave structure, they are absorbed by the material exponentially with distance and according to the dielectric properties and size of the sample. As a precaution in proper operation, a dummy water load is attached to load-end of the waveguide to absorb the residue of microwave energy that was not absorbed by the material. Provided that the load can move at a constant speed along the structure, each part of it would experience the same total field strength once steady-state conditions are achieved (i.e., after at least one length of material has already passed through the structure).

The example of the application of traveling wave applicator for processing polymeric materials can be found in the literature. A schematic set-up for a traveling wave applicator is depicted in Figure 1.20 (Chen et al., 1993).

The system consists of variable (0 to 120 W) microwave power generator operating at 2.45 GHz. Adapter that connects the coaxial cable to the waveguide. Samples were separated by Teflon sheets and placed into 3.3 cm high × 1.27 cm inside diameter cylindrical Teflon sample holder. The sample holder was placed at the maximum electric field region of the waveguide. An adapter is fastened to the other end to couple the coaxial cable to the waveguide for the traveling mode. The fiberoptic temperature sensor was inserted into the center of the sample to measure the sample temperature. Various starting temperatures for the investigated materials were induced by external heating of the waveguide by a strip heater. In such a case, the traveling wave applicator applies power from the generator into the waveguide, allowing absorption of the energy by one pass of the waves through the sample before being absorbed by a terminator. Traveling wave applicators can be operated empty without risk to the generator. The waves cannot be reflected back and forth; therefore, the electric field strength is relatively low (Chen et al., 1993).

Microwave Reactors

Specialized microwave reactors for chemical synthesis are commercially available from several companies:

Anthon-Paar (http://www.anthon-parr.com)
Biotage (http://www.biotage.com)
CEM (http://www.cem.com)
Ertec (http://www.ertec.pl)
Milestone (http://www.milestone.com)
Plazmatronika (http://www.plazmatronika.pl)

They offer a number of multimode and single-mode reactors that are mostly adopted from microwave multimode cavities and single-mode systems developed for digestion of analytical samples, respectively. The reactors can be equipped with built-in magnetic stirrer and direct temperature control by means of an infrared pyrometer, shielded thermocouple or fiberoptical temperature sensor, and continuous power feedback control, which enable heating of a reaction mixture to a desired temperature without thermal runaways. In some cases, it is possible to work under a reduced pressure or under pressurized conditions within the cavity or in reaction vessels. The comparison of different type of microwave reactors with a household microwave oven (i.e., Sharp R-220A) is given in Table 1.6 (Nuchter et al., 2003). However, this equipment is limited to laboratories due to lack of scalability; the manufacturers work to develop products that can increase the volume substantially.

In fact, the strict distinctions between multimode and single-mode systems can be misleading in laboratory practice. For instance, in single-mode cavities, the material is irradiated only from one side, and, in particular, for the systems with high power density (around 300 W/L), electrical as well as temperature in homogeneities are observed. Furthermore, vessels with diverse reaction mixtures inside the single-mode cavity influence the field geometry, and new modes of microwaves may be created by reflections and subsequent interferences, which eventually may result in a two- or three-mode system. Therefore, from a practical point of view, it seems to be better to follow the rule that batch reactions on a scale higher than 50 mL should be performed in multimode microwave reactors, while the reactions on a scale lower than 10 mL should be carried out in single-mode reactors. Unfortunately, the typical region for chemical syntheses performed in the laboratory is often in the range of 10 to 50 mL, so we have to leave to the experience of the chemist to assess the kind of materials in his or her hands and choose the type of microwave cavity. Even a small amount of the material that is a good microwave absorber can be very efficiently heated in single-mode as well as multimode cavities. Nevertheless, in multimode reactors (cavities), it is difficult to obtain high radiation intensities (power densities) within small samples of materials because usually the cavity is much larger than the sample. In such a case, only a small part of the microwave radiation is absorbed by the sample, while the rest of radiation is dissipated and converted into heat on the walls of the cavity. On the other hand, in single-mode cavities, the amount of the material is limited by the size of the cavities, which have to be smaller than one half-wavelength, which is about 6 cm for 2.45 GHz microwave frequency; otherwise, the samples are heated unevenly resulting in temperature inhomogeneities, particularly for systems with high average power densities.

The reader should be also aware that a number of commercially available systems for high-throughput as well as combinatorial chemistry have recently been provided by microwave equipment manufactures, but their description is out of the scope of this book. Extensive descriptions of such systems can be found in recently published books (Kappe et al., 2005; Loupy et al., 2006).

Temperature Monitoring

Every efficient application of microwave energy to perform chemical syntheses requires reliable temperature measurement as well as continuous power feedback control, which enable heating of reaction mixtures to a desired temperature without thermal runaways. Moreover, power feedback control systems that are operated in the most microwave reactors enable a synthesis to be carried out without knowing the dielectric properties and/or conductive properties of all the components of the reaction mixture in detail. On the other hand, temperature control during microwave irradiation is a major problem that one faces during microwave-assisted chemical reactions. Maintaining good thermal contact with the material being heated is crucial when heating using microwave irradiation, and it is important that temperature probes produce a minimum perturbation to the existing field in a microwave cavity. In general, temperature in microwave field can be measured by means of:

- Fiberoptic thermometer
- Shielded thermocouple
- Pyrometer (infrared sensor)

Fiberoptic thermometers can be applied up to 300 °C but are too fragile for real industrial applications. In turn, optical pyrometers and thermocouples can be used, but pyrometers measure only surface temperatures, which can be lower than the interior temperatures in reaction mixtures. Application of thermocouples, which in the case of microwaves are metallic probes, screened against microwaves, can result in arcing between the thermocouple shield and the cavity walls, leading to failures in thermocouple performance.

The temperature of microwave-irradiated samples can be also measured by inserting either a thermocouple or thermometer into the hot material immediately after turning off microwave power. This rather simple procedure may sometimes help if no other means of temperature measurements is provided. However, it must be stressed that the temperature of many materials can drop quickly as soon as microwave power is switched off (Bogdal, 2005).

Methods for Performing Reactions Under Microwave Irradiation

Two pioneering works for the synthesis under microwave conditions that were published almost 20 years ago described several organic syntheses that were completed in household microwave ovens with high yield when conducted in sealed vessels (Gedey et al., 1998; Giguere et al., 1986). They used commercially available screw-up pressure vessels made of glass, Teflon, and PTFE (i.e., being transparent for microwaves) (Fig. 1.21). This strategy has been successfully applied to a number of syntheses, but it always generates a risk of hazardous explosions. Later, there were reported 45 different reaction procedures with a commercial microwave oven and poly(ethylene terephthalate) vessels designed for acid digestion (Majetich et al., 1995). Since then, different techniques have been developed.

The simplest method for conducting microwave-assisted reactions involves irradiation of reactants in an open vessel. Such a method, termed *microwave-organic reaction enhancement* (MORE), was developed (Bose et al., 1991). During the reaction, reactants are heated by microwave irradiation in polar, high-boiling solvent so that the temperature of reaction mixture does not reach the boiling point of a solvent. Despite the convenience, a disadvantage of the MORE technique consists in limitation to high-boiling polar solvents such as DMSO, DMF, *N*-methylmorpholine, diglyme,

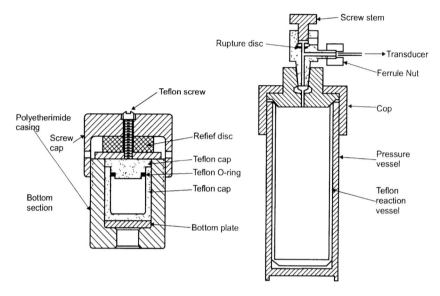

Fig. 1.21   Reaction vessels for performing chemical reactions under pressurized conditions. Reprinted from Mingos, D.M.P., Baghurst, D.R. 1991. Chem. Soc. Rev. 20:1, with permission.

etc. The approach has been adapted to lower-boiling solvents (e.g., toluene) (Morcuende et al., 1994); however, it still generates a potentially serious fire hazard. For reactions under reflux conditions, household microwave ovens have been modified by drilling a shielded opening to prevent leakage of microwave, and through which the reaction vessel has been connected to a condenser (Fig. 1.22) (Mingos et al., 1991).

Microwave heating has been proved to be of benefit particularly for the reactions under "dry" media in open vessel systems (i.e., in the absence of a solvent, on solid support with or without catalysts) (Bram et al., 1992). Reactions under "dry" conditions were originally developed in the late 1980s (Laszlo, 1987), but solventless systems under microwave irradiation offer several advantages. The absence of solvent reduces the risk of explosions when the reaction takes place in a closed vessel. Moreover, aprotic dipolar solvents with high boiling points are expensive and difficult to remove from the reaction mixtures. During microwave induction of reactions under dry conditions, the reactants adsorbed on the surface of alumina, silica gel, clay, and other mineral supports absorb microwaves, whereas the support does not, nor does it restrict the transmission of microwaves. Consequently, such supported reagents efficiently induce reactions under safe and simple conditions.

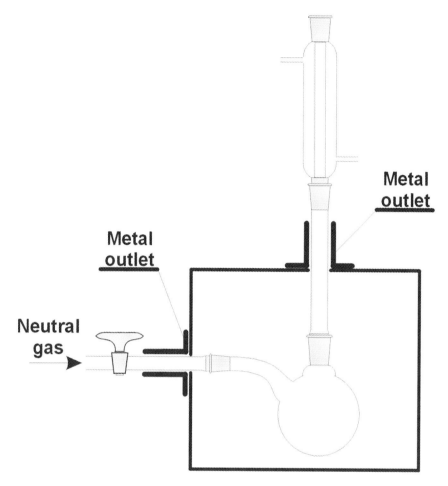

Fig. 1.22 Modification of a microwave cavity for performing chemical reactions under reflux conditions.

Finally, the mixture of neat reagents in an open vessel can lead to a reaction under microwave conditions provided that one of the reagents is liquid or a low melting solid "microwave materials technologies" that couples well with microwaves (Bogdal, 2005).

As will be shown in the next chapters, all of these methods adopted from organic synthesis have been successfully applied for polymer synthesis under microwave irradiation.

References

Bogdal, D. 2005. Microwave-Assisted Organic Synthesis, One Hundred Reaction Procedures. Elsevier. Amsterdam.

Bose, A.K., Manhas, M.S., Ghosh, M., Shah, M., Raju, V.S., Bari, S.S., Newaz, S.N., Banik, B.K., Chaudhary, A.G., Barakat, K.J. 1991. J. Org. Chem. 56:6968.

Bram, G., Loupy, A., Majdoub, M., Gutierrez, E., Ruiz-Hitzky, E. 1990. Tetrahedron 46:5167.

Bram, G., Loupy, A., Villemin, D. 1992. In: Solid Supports and Catalysts in Organic Chemistry. Smith, K. (Ed.). Ellis Harwood. London.

Chen, M., Siochi, E.J., Ward, T.C., McGrath, J.E. 1993. Polym. Eng. Sci. 33:1092.

Gabriel, C., Gabriel, S., Grant, E.H., Halstead, B.S.J., Mingos, D.M.P. 1998. Chem. Soc. Rev. 27:213.

Gedye, R.N., Smith, F.E., Westaway, K.C. 1988. Can. J. Chem. 66:17.

Giguere, R.J, Bray, T.L., Duncan, S.M., Majetich, G. 1986. Tetrahedron Lett. 27:4945.

Kaatze, U. 1995. Radiat. Phys. Chem. 45:539.

Kappe, O., Stadler, A. 2005. Microwave in Organic and Medical Chemistry. Wiley-VCH. Weinheim.

Kubel, E. 2005. Ind. Heating January: 43-53.

Laszlo, P. (ed). 1987. Preparative Chemistry Using Supported Reagents. Academic Press. New York.

Loupy, A. 2006. Microwaves in Organic Synthesis. Wiley-VCH, Weinheim.

Majetich, G., Hicks, R. 1995. Radiat. Phys. Chem. 45:567.

Metaxas, A.C., Meredith, R.J. 1983. Industrial Microwave Heating. Peter Perigrinus. London.

Mingos, D.M.P., Baghurst, D.R. 1991. Chem. Soc. Rev. 20:1.

Morcuende, A., Valverde, S., Harradon, B. 1994. Synlett 89.

National Academy Press. 1994. Microwave Processing of Materials. National Academy Press.

Nuchter, M., Muller, U., Ondruschka, B., Tied, A., Lautenschlanger, W. 2003. Chem. Eng. Technol. 26:1207.

Stuerga, D., Delmotte, M. 2003. Wave-material interaction, microwave technology and equipment. In: Microwaves in Organic Synthesis. Loupy, A. (Ed.). Wiley-VCH. Weinheim.

Thuery, J. 1992. Microwaves: Industrial, Scientific, and Medical Applications. Artech House. Boston/London.

2

OVERVIEW OF POLYMERIZATION PROCESSES UNDER MICROWAVE CONDITIONS IN COMPARISON WITH CONVENTIONAL CONDITIONS

The purpose of this chapter is to provide useful details concerning the application of microwave irradiation to the polymer chemistry. Research in this area has shown potential advantages of microwaves to not only drive the polymerization reaction but also to perform them on a reduced time scale. In some cases, afforded products exhibited properties that may not be possible using conventional thermal treatments. In this chapter, some examples of the polymerization reactions, which are divided according to the polymerization methods used during the reactions, are described to show that bulk polymerization reaction are best suited for microwave conditions, as mentioned in the previous chapter, but also to show that suspension, emulsion, and polymerization in a fluid bed can be carried out under microwave irradiation. In most of these cases, special equipment was built to apply microwave irradiation.

In the next chapters, detailed descriptions of many polymerization reactions and cross-linking, as well as processing under microwave conditions, are presented, but unlike in this chapter, microwave reactors, or at least a modified household microwave oven, were used.

Because the effects of microwave irradiation on chemical reactions are generally evaluated by comparing the time needed to obtain a desired yield of final products with respect to conventional thermal heating, for this purpose, the syntheses run under microwave conditions were compared with conventional heating methods. Moreover, the shape and size of the reaction vessel are important factors for the processing of material under microwave irradiation as well as applied microwave system (i.e., applicator and temperature detection). Therefore, in each case we have tried to briefly describe microwave systems that were used by different research groups together with amounts of reagents to give readers a deeper feeling about the microwave experiments.

However, a brief overview of the history of the application of microwaves in polymer chemistry and characteristics of dielectric properties of polymers is given first.

Brief Story of the Application of Microwaves in Polymer Chemistry

Since the late 1960s, microwave irradiation has been used to synthesize and process polymeric materials as rapid heating and melting of neat and mineral-filled plastics for a number of purposes such as fast curing of thermosetting resins and composites, polymerization of vinyl monomers, rapid drying of aqueous solutions or dispersion of polymers and resins, and heat drawing of polymer rods and tubings (Parodi, 1998). For example, microwave irradiation was applied to cure of epoxy resins (Williams, 1968), and the use of microwave irradiation for bulk polymerization of dental materials was also reported (Nishii, 1968). It allowed the obtainment of porosity-free acrylic resin specimens of similar physical and mechanical properties as the water bath cured specimens. In turn, emulsion polymerization of various vinyl monomers (i.e., styrene, acrylic and methacrylic esters, acrylic and methacrylic acids) was described for the radiofrequency and microwave frequency range (Vanderhoff, 1969).

A remarkable achievement of the late 1960s was the application of microwave irradiation for the continuous vulcanization of extruder rubbers and the discontinuous vulcanization of molded rubber articles (Akyel et al., 1989; Krieger, 1992, 1994). The microwave-assisted vulcanization of rubber compounds that was used industrially since the 1970s is the most important application of microwave heating to polymeric materials in terms of number of installed plants. For example, in 1992, the microwave vulcanization of extruded rubber weather stripping was used commercially in more than 600 automotive and construction industries worldwide (Krieger, 1992). Microwave processing offered rubber processors significant advantages over conventional processing, including improved product uniformity, reduced extrusion-line length, reduced scrap, and improved cleanliness and environmental sustainability compared with steam autoclaves, hot air, slat batch, or fluid bed heating processes (National Academy Press, 1994).

In the 1970s, the technology of microwave heat-drawing was developed for the manufacture of high-modulus polymeric materials. Many other applications of microwave heating were introduced, including the rapid bonding of polymeric films, sheets, and various components and fast curing of thermosetting resins. In the 1980s, microwave technology revealed almost simultaneously in polymer and organic chemistry an outstanding potential to cause strong acceleration and/or selective promotion of a number of chemical processes (Parodi, 1996).

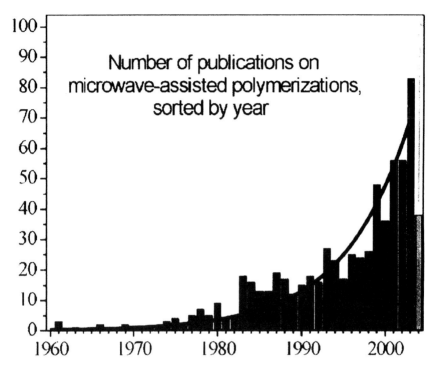

Fig. 2.1. Number of publications on microwave-assisted polymerizations. Reprinted from Wiesbrock, F., Hoogenboom, R., Schubert, U.S. 2004. Macromol. Rapid. Commun. 25:1739, with permission.

Recently, microwave technology applied to polymer processing, and particularly to the polymerization process, has become a subject of a great deal of academic and industrial research with much scientific and patent literature being generated (Bogdal et al., 2003, 2006a; Parodi, 1996; Wiesbrock et al., 2004). Microwave technology has become widely accepted and popular, with unconventional technology in polymer chemistry as an alternative to, and often an improvement on, conventional heating. This is clearly evident from the annual number of publications on microwave-assisted polymer chemistry, growing rapidly with more than 600 publications by the end of 2004 (Fig. 2.1) (Wiesbrock et al., 2004).

Characterization of Dielectric Properties of Polymers

As stated in the previous chapter, the reorientation of dipoles in the electric field is the principal mechanism of microwave absorption of materials

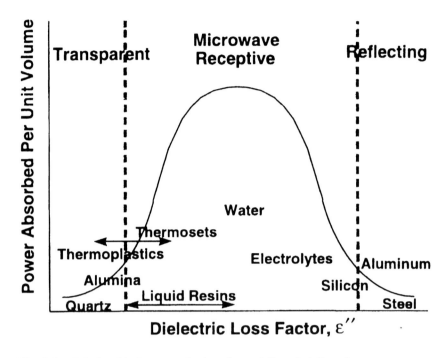

Fig. 2.2.   Relationship between the loss factor (ε'') and ability of some common materials to absorb microwave energy. Reprinted from Thostenson, E.T., Chou, T.W. 1999. Composites Part A 30:1055, with permission.

as well as polymers. The materials with the greatest dipole mobilities will exhibit the most effective coupling. Microwaves will couple most efficiently with the strongest dipole in a system and has the potential to selectively heat polymers in mixtures. The properties of materials under microwave irradiation (i.e., in highly oscillating electric fields) are characterized by their dielectric properties such as the complex permeability ($\varepsilon^* = \varepsilon' - j\,\varepsilon''$). The real permeability (ε') component characterizes the ability of the material to be polarized by the electric field and thus the ability of microwaves to propagate into the material. The imaginary part of the complex electric permeability (ε'') is usually called the loss factor and indicates the ability of the material to dissipate the energy (i.e., the efficiency of conversion of electromagnetic radiation into heat; see Chapter 1). The relationship between the loss factor (ε'') and ability of some common materials to absorb microwave energy is shown in Figure 2.2 (Thostenson et al., 1999).

Materials with a high conductance and low capacitance (such as metals) have a high loss factor (ε'') and the penetration depth approaches zero. Thus, these materials are considered to be reflectors of microwave irradiation. In opposite, materials with low values of the loss factor (ε'') have a large penetration depth. As a result, very little energy is absorbed in the material, and it is transparent to microwave irradiation like quartz glass. Because of such behavior, microwaves transfer energy most effectively to materials that posses the lost factor (ε'') in the middle of conductivity range (Fig. 2.2). In contrast, conventional heating transfers heat most efficiently to materials with high conductivity (Thostenson et al., 1999). It is worth stressing that most of the polymers (i.e., thermoplastics, thermosets, and liquid resins) are in the microwave receptive region for the given frequency (Fig. 2.2); however, a strict distinction between nonpolar and polar polymers (i.e., polymers with nonpolar pendant groups and polar pendant groups, respectively) has to be kept in mind.

The real permeability (ε') and the loss factor (ε'') for nonpolar polymers at room temperature for the frequency range of 10^4 to 10^9–10^{10} Hz are shown in Table 2.1 (Parodi, 1996).

As expected, nonpolar polymers display no or negligible decrease of real permeability (ε') and always very low values of the loss factor (ε'') (of the order of 10^{-3}). Similar properties are also shown by polystyrene: real permeability (ε') decreasing from 2.55 to 2.49, and maximum of the loss factor (ε'') approximately 4×10^{-3} from 1 kHz to 1 GHz (Parodi, 1996).

The real permeability (ε') and the loss factor (ε'') values at room temperature for some important polar polymers over the similar frequency range of 10^3 to 10^9 Hz are collected in Table 2.2. This shows the significantly higher dielectric loss capacities with the real permeability (ε') values at least one order of magnitude higher than those for nonpolar polymers. Such data are particularly exemplified through the loss factor (ε'') values typically 100-fold higher over the entire region (Parodi, 1996).

Moreover, polymer dielectric properties can vary during a processing cycle or if a phase change occurs as temperature varies, solvent is removed, and the reaction proceeds changing the type and concentration of dipoles. Generally, several distinct dielectric relaxation processes are present in solid polymeric materials as is shown in Figure 2.3, which is a scan of the loss factor (ε'') at constant frequency as a function of temperature (Parodi, 1996).

The multiplicity of dielectric relaxation (α, β, γ) is associated with the different possible modes of orientational polarization and related relaxation phenomena occurring in polymer matrices. The α relaxation at the lowest frequencies is associated with the glass transition temperature of the polymer (i.e., the movement of the main chain via cooperative motions of a multiplicity of chain segments). The weaker relaxations (i.e., β, γ, etc.) appearing at progressively higher frequencies and lower temperatures are associated with motion of signal structural elements like rotations of dipolar side groups, conformational changes of side groups, and local motions of

Table 2.1. Dielectric properties of some nonpolar polymers at radio and microwave frequencies at room temperature (20–25 °C).

Polymer	Frequency (Hz)							
	10^4	10^5	10^6	10^7	10^8	10^9	3×10^9	10^{10}
Polypropylene*¶								
ε'	2.4	—	2.4	—	—	1.2	—	—
$\varepsilon'' \times 10^3$	<1.2	—	<1.2	—	—	4.5	—	—
Butyl rubber†**								
ε'	2.37	2.36	2.35	2.35	2.35	—	2.35	—
$\varepsilon'' \times 10^3$	6.4	3.4	2.4	2.4	2.4	—	2.4	—
Butadiene-styrene rubber‡**								
ε'	2.5	2.5	2.5	2.45	2.45	—	2.45	—
$\varepsilon'' \times 10^3$	2.5	4.5	9.5	16.9	17.4	—	10.8	—
Hevea rubber§**								
ε'	2.4	2.4	2.4	2.4	2.4	—	2.15	—
$\varepsilon'' \times 10^3$	2.4	3.4	4.3	7.7	12.0	—	6.4	—
FEP‖¶								
ε'	—	2.1	—	2.1	—	2.1	—	2.1
$\varepsilon'' \times 10^3$	—	0.5	—	≈1	—	2.3	—	0.5

*Average value for semicrystalline, isotactic polymer (crystallinity (60%).
†Isobutene-isoprene copolymer (99:1–98:2).
‡Butadiene-styrene copolymer (75:25 wt/wt).
§Purified, natural rubber (pale crepe): >98% *cis*-1,4-polyisoprene.
‖Tetrafluoroethylene-hexafluoropropene copolymer (90:10–85:15 wt/tw).
¶Data from Brandrup, J., Immergut, E.H. (Eds.). 1989. Polymer Handbook, 3rd. ed. Wiley. New York.
**Data from Tinga, W.R., Nelson, S.O. 1973. Microwave Power 8:23.
Reprinted from Parodi, F. 1996. Physics and chemistry of microwave processing. In: Comprehensive Polymer Science. 2nd Supplement. Aggarwal, S.L., Russo, E. (Eds.). Pergamon. Oxford:669, with permission.

individual segments of the main chain and do not involve of segmental mobility of the whole polymer backbone (Parodi, 1996).

Regarding different polymerization methods, under microwave conditions polymers were prepared by bulk and solution polymerization as well as suspension and emulsion polymerization methods. Because most cases of polymerization under microwave irradiation are related to bulk and solution polymerization, in this chapter we decided to discuss the suspension as well as emulsion polymerization reactions. Another reason to choose these sets of experiments was that different equipment was used during these reactions compared with most reports; the reader can realize that

Table 2.2. Dielectric properties of some polar polymers at radio and microwave frequencies at room temperature (23–26 °C).

Polymer	Frequency (Hz)							
	10^3	10^4	10^5	10^6	10^7	10^8	10^9	10^{10}
Polychloroprene*								
ε′	—	6.2	6.1	5.7	4.7	3.4	2.84†	—
ε″	—	0.205	0.183	0.541	0.940	0.544	0.136†	—
PMMA‡								
ε′	3.2	3.0	2.9	2.8	2.75	2.7	—	2.57§††
ε″	0.140	0.100	0.082	0.070	0.048	0.032	—	0.018§††
PET‖††								
ε′	3.25	—	—	3.0	—	—	2.8	—
ε″	0.016	—	—	0.048	—	—	0.008	—
Nylon-6¶††								
ε′	3.5	3.4	3.4	3.3	3.2	3.2	3.1	—
ε″	0.035	0.048	0.065	0.079	0.096	0.096	0.062	—
Cellulose**††								
ε′	7.2	7.0	—	—	5.8	5.6	—	—
ε″	0.145	0.21	—	—	0.46	0.39	—	—
PVDF‡‡								
ε′	7.7–13.2††	—	—	6.0–7.6††	—	3.3§§	2.9§§	2.6§§
ε″	0.10–0.25††	—	—	0.9–1.76††	—	0.54§§	0.3§§	0.09§§

*1,4-Poly(2-chloro-1,3-butadiene), essentially amorphous. Data from Tinga, W.R., Nelson, S.O. 1973. Microwave Power 8:23.

†Values at 3 GHz.

‡Poly(methylmetacrylate), atactic (amorphous), data from Broadhurst, M.G., Bur, A.J. 1965. J. Res. Natl. Stand. 69C:165.

§Values at 3 GHz.

‖Poly(ethyleneterephthalate) semicrystalline: Mylar A (Du Pont) biaxially oriented film.

¶Semicrystalline, injection-molded specimens, values at 30 °C.

**Pure, dry, values at 20 °C.

††Data from Brandrup, J., Immergut, E.H. (Eds.). Polymer Handbook, 3rd. ed. Wiley. New York. 1989.

‡‡Poly(vinylidene fluoride), semicrystalline. Due to its ferroelectric characteristics, ε′ and ε″ may vary within intervals depending on its previous electricalization and thermal history.

§§Data from Wentink, T. 1961. J. Appl. Phys. 32:1063.

Reprinted from Parodi, F. 1996. Physics and chemistry of microwave processing. In: Comprehensive Polymer Science. 2nd Supplement. Aggarwal, S.L., Russo, E. (Eds.). Pergamon. Oxford.669, with permission.

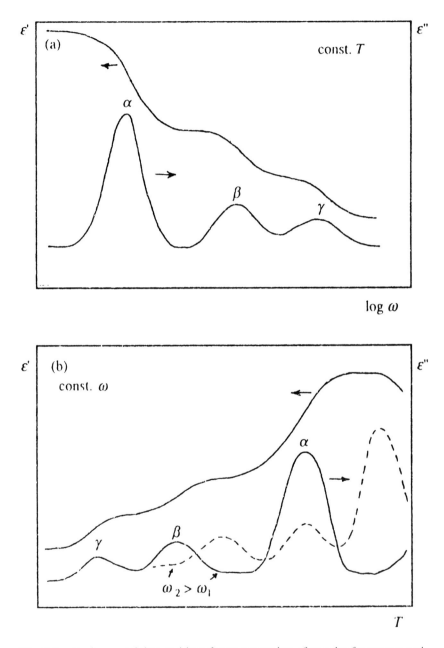

Fig. 2.3. Real permeability and loss factor versus log of angular frequency and temperature. Reprinted from Parodi, F. 1996. Physics and chemistry of microwave processing. In. Comprehensive Polymer Science. 2nd Supplement. Aggarwal, S.L., Russo, E. (Eds.). Pergamon. Oxford:669, with permission.

there is no limitation to commercially available microwave reactors while performing reactions under microwave conditions.

Temperature Control

Like most reactions carried out under microwave conditions, one of the major problems in the case of microwave-assisted polymerization is to achieve accurate temperature control, because, as mentioned in Chapter 1, conventional thermometers and thermocouples cannot be used in a microwave field. Another problem is related to inhomogeneous temperature distribution in the polymerization medium and temperature gradient that usually develops with higher temperatures in the center of the reactor and decreases toward the reactor walls. Discussion on the development of thermal gradients recorded by means of a thermovision camera during the bulk synthesis of high-molecular-weight epoxy resins is provided in Chapter 4 of Bogdal et al. (2004), while examples of thermal gradients during the reactions on mineral support were more recently published (Bogdal et al., 2006b). It was found that a proper temperature measurement in the case of heterogeneous reaction mixture is very difficult. In order to maintain a good temperature homogeneity and make some comparison with the experiments under conventional conditions, an effective stirring has to be provided—perhaps together with a small amount of an inert solvent.

Another important factor is that there is a general agreement that the application of fiberoptic thermometers is the reliable way to determine temperature under microwave conditions. Applying the thermovision camera, it was found that for the reactions in heterogeneous systems under microwave irradiation, the temperature measurement with a fiberoptic thermometer can lead to serious errors like pyrometry—in particular, for those experiments that are planned without any attention paid to temperature homogeneity of the reaction mixture. In the latter case, a high-temperature gradient within the reaction mixture generated by microwaves leads to higher conversion of reactants and/or reaction rates, which in turn might be a reasonable explanation to so-called nonthermal microwave effects (i.e., an increase in reaction rates that is inadequate for the temperature of reaction medium). Therefore, before considering the increase in reaction rates by special microwave effects (thermal or nonthermal), it is necessary to consider all the factors that might influence chemical reactions under microwave conditions, like a reaction mechanism, temperature profiles (gradients), and, in particular, proper design of experiments (Bogdal et al., 2006b).

In order to avoid such problems, a system with a special temperature control and effective stirring can be designed (Fig. 2.4); however, it is not typical for most equipment mentioned in this book (Albert et al., 1996).

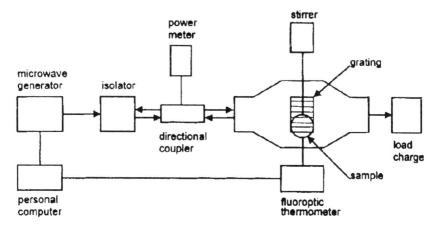

Fig. 2.4.   Schematic diagram of the microwave polymerization reactor system. Reprinted from Albert, P., Holderle, M., Mulhaupt, R., Janda, R. 1996. Acta Polym. 47:74, with permission.

A microwave generator produces microwaves in the range of 50 to 850 W (continuous) with a frequency of 2.45 GHz. The microwaves are forced to propagate inside the waveguide with the TE_{10} mode, which is not disturbed through the overdimensionalized rectangular waveguide (applicator). It is equipped with a grating that enables a visible control from outside. Also, it contains a mode filter that erases probable new modes that can be created in the sample. The reaction mixture is in a two-necked flask equipped with a stirrer made of Teflon, a fiberoptic temperature sensor, and a Teflon tube as argon inlet. The magnetron is protected from reflected microwaves by an isolator, and the power of the microwaves can be measured using a power meter. Eventually, the microwaves are absorbed in the water load so that no standing waves could exist. The temperature is registered with a computer and controlled through the magnetron by a backloop.

Suspension Polymerization

The microwave system was applied for the investigation of free radical suspension polymerization of methylmethacrylate (MMA) in an *n*-heptane solution in the presence of polystyrene-*block*-poly(ethene-*alt*-propene) (SEP) as a dispersing agent afforded polymethylmethacrylate (PMMA) samples with similar molecular weights and polydispersity under both conventional and microwave conditions (Albert et al., 1996).

Because *n*-heptane does not contain polar groups, it is nearly impossi-

Table 2.3. Efficiency of microwave irradiation at power input in 500 g of *n*-hexane and 2-propanol.

Solvent or mixture of solvents	Efficiency (%)
n-Hexane, pure	8
n-Hexane/2-propanol 9:1	50
n-Hexane/2-propanol 8:2	≥90
n-Hexane/2-propanol 7:3	≥90
n-Hexane/2-propanol 5:5	≥90
2-Propanol, pure	≥90

Reprinted from: Nuchter M., Muller, U., Ondruschka, B., Tied, A., Lauten-schlanger, W. (2003) Chem. Eng. Technol. 26:1207, with permission.

ble to heat *n*-heptane via dielectric heating with microwaves at a frequency of 2.45 GHz. If polar molecules are dissolved in *n*-heptane, they absorb microwaves and lead to much higher dielectric heating, which results in a higher temperature increase for a given microwave power. For example, the efficiency of microwave absorption (i.e., amount of the microwave energy absorbed by a solution compared to the entire microwave energy emitted to the reactor cavity) was measured for *n*-hexane/2-propanol mixtures (Table 2.3).

While pure *n*-hexane, which like *n*-heptane, absorbs only a little microwave energy, a mixture of 90 wt% *n*-hexane and 10 wt% 2-propanol clearly shows a significant effect. Moreover, a mixture of 80 wt% *n*-hexane and 20 wt% 2-propanol absorbs microwave radiation with an efficiency that is comparable to that of pure 2-propanol. Additional increases in the amount of 2-propanol does not lead to any further increase in efficiency (Nuchter et al., 2003). Similarly, MMA molecules, which played the same role as 2-propanol, lag behind the electric field which leads to a dielectric loss and heating. The heat dissipated very quickly to the neighboring molecules (*n*-heptane), and this resulted in a higher temperature of the solution.

The reactions were run for 1 hour at 70 °C with different monomer (9.0 to 28.3 vol%), SEP (21.7 to 5.4 wt%), and azobisisobutyronitrile (AIBN) (1.0 to 0.27 wt%) concentrations (Table 2.4). In a typical experiment, 30 mL of the reaction mixture was fed into a 50 mL reaction vessel that was equipped with a stirrer (200 rpm), fiberoptic thermometer, and gas inlet tube to avoid air in the reaction mixture. Then the reaction vessel was placed into a microwave cavity.

Temperature and power profiles for the investigated polymerization reaction are shown in Figure 2.5 and are typical for a great number of chem-

Table 2.4. Thermal (T) and microwave-induced (M) nonaqueous free radical dispersion polymerization of MMA in the presence of SEP dispersing agent.

Method of activation	n-Heptane (vol%)	MMA (vol%)	SEP* (vol%)	AIBN* (wt%)	Conversion (%)	Diameter† (nm)
T1	91	9	21.7	1	42	67 ± 16
M1	91	9	21.7	1	50	92 ± 53
T2	83.5	16.5	10.8	0.53	73	144 ± 23
M2	83.5	16.5	10.8	0.53	71	134 ± 42
T3	71.7	28.3	5.4	0.27	75	1330 ± 133
M3	71.7	28.3	5.4	0.27	73	1370 ± 107

*Weight percent with respect to MMA.
†Average diameter of the particles with standard deviation.
Reprinted from Albert, P., Holderle, M., Mulhaupt, R., Janda, R. 1996. Acta Polym. 47:74, with permission.

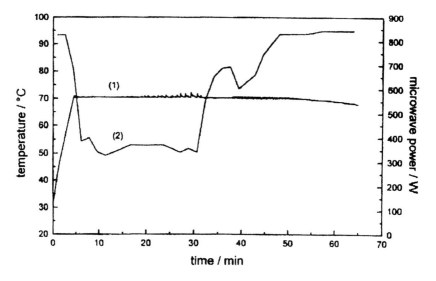

Fig. 2.5. Typical profiles of temperature (1) and power (2) versus time in the traveling waveguide during the polymerization of sample M3 (28.3 vol% MMA). Reprinted from Albert, P., Holderle, M., Mulhaupt, R., Janda, R. 1996. Acta Polym. 47:74, with permission.

ical reactions under microwave conditions. The figure (Fig. 2.5) displays the average microwave power (2) and the resulting temperature (1) for run M3 (28.3 vol% MMA) (Table 2.4). As soon as the temperature of 70 °C is reached, the temperature program reduces the microwave power to a medium value of 350 W. When MMA was converted into PMMA, which ab-

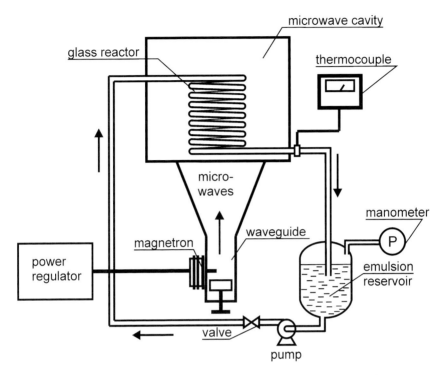

Fig. 2.6. A bench-scale microwave polymerization reactor. Reprinted from Correa, R., Gonzalez, G., Dougar, V. 1998. Polymer 39:1471, with permission.

sorbed less microwave energy than MMA, more microwave power was required (e.g., the microwave power is increased after 30 minutes from 350 W to 840 W after 48 minutes).

However, under such conditions, no significant effect of microwave irradiation was detected, and molecular weights, polydispersities, stereoregularities, particle size, and their distribution were similar for both activation methods.

Emulsion Polymerization

Conversely, a more recent study in which a bench-scale microwave polymerization reactor was used (Fig. 2.6) reported that emulsion polymerization of styrene could be carried out more rapidly with significant saving of energy and time compared with conventional methods (Correa et al., 1998).

For all the experiments, the concentration range of materials was as follows: potassium persulfate KPS (initiator) 0.04 to 0.16 wt%, sodium lauryl

Table 2.5. Experimental conditions for samples with different amounts of initiator.

	A	B	C	D
Initiator (%)	0.04	0.08	0.12	0.16
Emulsifier (%)	1.24	1.24	1.24	1.24
Monome:water*	1:3	1:3	1:3	1:3
Power (W)	800	800	800	800
Power on (sec)	20	20	20	20
Power off (sec)	600	600	600	600
Total time† (sec)	620	620	620	620

*Volume ratio.
†Time for each cycle.
Reprinted from Correa, R., Gonzalez, G., Dougar, V. 1998. Polymer 39:1471, with permission.

sulfate (emulsifier) 1.24%, and monomer-to-water ratio 1:3, 1:16, and 2:3. The reaction temperature for both methods was 70 °C, while two power levels (175 W and 800 W) of microwave irradiation were applied (Table 2.5). In order to reduce the risk of hazardous explosions associated with prolonged irradiation time, a pulse microwave irradiation was applied.

The reaction time for the conventional heating method was found to be about 70 times longer than that for the microwave irradiation method. Moreover, it was observed that the molecular weight of the polystyrene samples prepared by microwave irradiation (M_W = 350,000 g/mol) was 1.2 times higher than that of polystyrene obtained by the conventional heating. As it can be expected, the requested period of time under microwave irradiation depended on the applied microwave power and was longer for lower power level (Correa et al., 1998). In fact, microwave power was used in turn on/off cycles of 20 and 600 seconds, respectively, while the reaction time was recorded only during turn on cycles. In our opinion, both turn on and turn off cycles had to be counted, which means that the total reaction time under microwave irradiation was comparable with the reaction time under conventional conditions.

In turn, emulsion polymerization of styrene under microwave irradiation in the presence of KPS as an initiator was described in the next report (Palacios et al., 1996). The reactions were carried out in glass tubes of 20-mL capacity, while 15 mL of the reaction mixture with different initiator concentration (0.2 to 31.1 mmol/L) (Table 2.6) was introduced into the tubes. During the experiments, each tube was placed in a 2-L glass beaker filled with vermiculite. The beakers were irradiated in a multimode microwave with 389 W power in the magnetron, which was turned on and off twice per minute to maintain a constant temperature (50 °C).

Table 2.6. Emulsion polymerization of styrene initiated by microwaves.

No.	KPS initiator concentration [(mol/L) × 10⁴]	Polymer yield (%)†	M_n [× 10⁻³ (g/mol)]	M_w/ M_n	R_p [(mol/L × sec) × 10³]	Time required for conversion of 50% (sec)
1	0	5.0	—	—	0	—
2	2.90	52.0	385.1	3.23	3.57	1100
3	9.70	73.0	—	—	6.20	800
4	31.15	80.0	131.0	2.51	7.88	830
5	124.6	91.0	79.7	3.35	9.65	350
6	192.6	97.0	74.4	2.38	13.63	240
7	269.6	95.0	10.8	3.52	13.65	245
8	311.5	66.0	—	—	13.75	550
9*	192.6	69.0	70.1	2.41	51.84	18,500

*(6 hours), conductive heating. T = 50 °C.
†Time = 33 minutes.
R_p, polymerization rate.
Reprinted from Palacios, J., Velverde, C. 1996. N. Polym. Mater. 5:93., with permission.

In order to reach a conversion of styrene of 70%, constant heating in an oil bath for as long as 6 hours was required in comparison with only 8.3 minutes in the microwave oven. Calculated values of R_p (mol/L s) for both processes showed that the rate of microwave-assisted emulsion polymerization of styrene was more than 26 times higher than the reaction activated by conventional heating (Fig. 2.7) (Palacios et al., 1996).

Molecular weight distribution and polydispersities of the polystyrene samples obtained under microwave and conventional conditions were similar for both activation methods and depended on the initiator concentration (Fig. 2.8) (Palacios et al., 1996).

Recently, emulsion polymerization of MMA under pulsed microwave irradiation was studied (Zhu et al., 2003). The reactions were carried out in the self-designed single-mode microwave reaction apparatus with frequency of 1250 MHz and its pulse width of 1.5 or 3.5 μsec (Fig. 2.9).

The output peak pulse power, duty cycles, and mean of output power were continuously adjustable within the range of 20 to 350 kW, 0.1% to 0.2%, and 2 to 350 W, respectively. Temperature during microwave experiments was maintained by immersing the reaction flask in a thermostated jacket with a thermostatic medium with little microwave absorption (i.e., tetrachloroethylene). In a typical experiment, 8.0 mL of MMA, 20 mL of deionized water, and 0.2 g of sodium dodecyl sulfonate were transferred

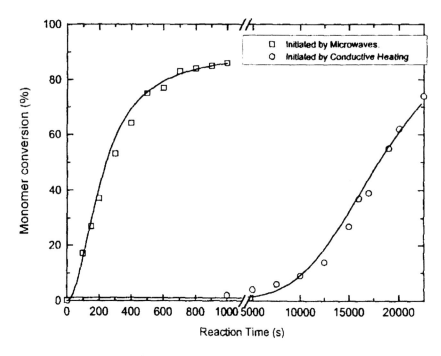

Fig. 2.7.   Conversion of microwave-assisted and conventional emulsion polymerization of styrene versus time. Reprinted from Palacios, J., Velverde, C. 1996. New Polym. Mat. 5:93, with permission.

into a 100-mL reaction flask and placed in the microwave cavity. When the temperature reached a preset value, 10 mL of aqueous solution of initiator (i.e., KPS) was added, and the flask was exposed to microwave irradiation. The results obtained under microwave irradiation were compared with those from conventional experiments. For the microwave experiments, the amount of initiator used to reach constant conversion was reduced by 50% at the same polymerization rate, but at the same initiator concentration, 0.15 and 0.20 wt%, the polymerization rate increased by a factor of 131% and 163%, respectively.

In order to reveal the cause of accelerating polymerization by microwaves, the decomposition rate constant k_d of KPS was determined under microwave irradiation. As shown in Figure 2.10, at temperature of 68 °C, the k_d value calculated from the slope was 8.05×10^{-5} sec^{-1}. At a similar temperature (70 °C) under conventional conditions, it was reported to be 3.35×10^{-5} sec^{-1} (Schumb et al., 1940). A comparison of two results showed that although the temperature under microwave conditions was 1.5 °C lower than that under conventional conditions, the k_d value under microwave con-

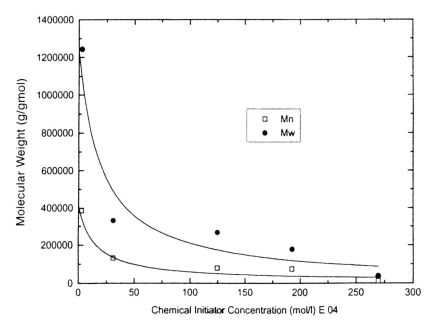

Fig. 2.8.   Molecular weight of the polystyrene samples obtained with different initiator concentration. Reprinted from Palacios, J., Velverde, C. 1996. New Polym. Mat. 5:93, with permission.

ditions increased by 2.4 times. In our opinion, it is a pity that the decomposition rate of KPS was not measured by the authors and is given from the literature, because the differences between the results under both conventional and microwave conditions can be caused by not using exactly the same set of reaction conditions (i.e., concentration, surfactant, etc.).

Because at a peak pulse power of 20 kW and duty cycle of 0.01%, the mean power of microwave irradiation was only 2 W, and the thermal effect of microwave is supposed to be very little; the authors suggested that it was the influence of so-called nonthermal microwave effect that played a main role in enhancing the initiator activity, raising its decomposition rate, and further accelerating the polymerization process. Similarly, at a temperature of 60 °C, the k_d that was 2.37×10^{-5} sec^{-1} under microwave conditions (Fig. 2.10) was larger than that under conventional conditions, which was reported to be 8.03×10^{-6} sec^{-1} (Schumb et al., 1940).

The molecular weight of polymer was 1.1 to 2.0 times higher than that obtained after conventional conditions. The glass transition temperatures (T_g), polydispersity index (PDI), and regularity of the polymers obtained using two processes (microwave versus conventional) were similar, indicating same polymerization mechanism.

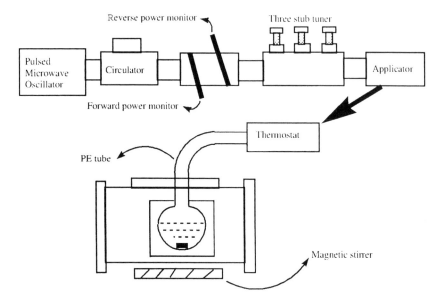

Fig. 2.9.   Apparatus for the pulsed microwave treatment of single-mode cavity. Reprinted from Zhu, X., Chen, J., Zhou, N., Cheng, Z., Lu, J. 2003. Eur. Polym. J. 39:1187, with permission.

Acceleration of Polymerization Reactions Under Microwave Conditions

As was mentioned earlier, the application of microwaves in chemistry was so attractive that from the very beginning it was realized that a number of chemical processes can be carried out with a substantial reduction in the reaction time in comparison to conventional processes. Reactions that usually take many hours or days can, under influence of microwave irradiation, be run in a considerably shorter time of several minutes or even seconds (Loupy, 2006). This phenomenon is not fully understood yet; however, there are two groups of theories that are proposed to explain the reduction of the reaction time under microwave conditions in comparison with processes under conventional conditions. According to the first group of theories, despite the fact that the course of chemical processes under microwave conditions is considerably shorter than under conventional conditions, the kinetics and mechanism of the reactions are still the same. The reduction of the reaction time is the result of sudden and, sometimes, uncontrollable temperature growth of the reaction mixture under microwave irradiation, which in turn leads to the increase in reaction rates following common kinetic laws. The second group of theories supposes that during

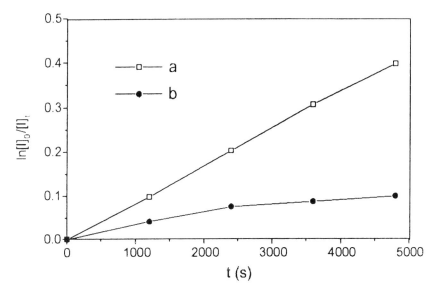

Fig. 2.10. Decomposition kinetics of KPS under PMI: (**a**) power = 20 kW, pulse width = 1.5 μsec, duty cycle = 0.01%, temperature = 68.5 ± 1 °C; (**b**) power = 60 kW, pulse width = 1.5 μsec, duty cycle = 0.02%, temperature = 60 ± 2 °C. Reprinted from Zhu, X., Chen, J., Zhou, N., Cheng, Z., Lu, J. 2003. Eur. Polym. J. 39:1187, with permission.

microwave irradiation of the reaction mixture, there is a specific effect of microwave activation that causes an increase in the reaction rates for which the bulk temperature of the reaction mixture is inadequate to explain. Such an effect has been accepted to be called the nonthermal microwave effect or the specific microwave effect (Bogdal, 2005). Recent critical reviews concerned with both group of theories can be found in the literature (de la Hoz et al., 2005; Nuchter et al., 2004; Perreux et al., 2001), respectively.

Recently, it has been postulated that the increase of the polarity of the reaction system (i.e., either development or increase of dipole moment) from the ground state (GS) toward the transition state (TS) can result in an acceleration of the reaction due to a stronger interaction of microwaves with the reagents during the course of the reaction. Thus, nonthermal microwave effect can be expected for the reaction with polar mechanisms when the stabilization of the TS is more effective than that of the GS, which results in an enhancement of the reactivity as a result of the decrease in activation energy (Fig. 2.11) (Perreux et al., 2006).

The most representative examples of this are concerned with unimolecular or bimolecular reactions between neutral molecules (i.e., dipole mo-

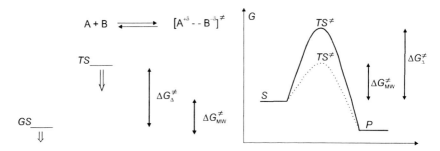

Fig. 2.11. Relative stabilization of a more polar transition state (TS) compared with the ground state (GS). Reprinted from Perreux, L., Loupy, A. 2006. Nonthermal effects of microwaves in organic synthesis. In: Microwaves in Organic Synthesis, A. Loupy (Ed.). Wiley-VCH, Weinheim, with permission.

ments are developed in the TS) and anionic reactions of tight ion pairs (i.e., charge localized anions), which leads to ionic dissociation in the TS (Perreux et al., 2006).

However, such effects can be masked or limited by either protic (e.g., alcohols) or aprotic solvents (e.g. DMF, CH_3CN, DMSO) that strongly interact with and absorb of microwaves. In such cases, the main interaction occurs between microwaves and polar molecules of the solvent that is used in an excess. Thus, particularly for samples large in size, microwaves do not penetrate the whole amount of the reaction mixture and a great amount of microwave energy is absorbed by the outer layers of the mixture (e.g., the penetration depth of water is approximately 1 to 2.5 cm). In turn, energy transport to the internal parts of the reaction mixture is via the classic heat conduction mechanism from the outer layers, which is commonly observed for conventional heating. Therefore, reaction rates under microwave conditions in a polar solvent solution should be nearly the same as those observed under conventional conditions. However, this can lead to the conclusion that the most favorable systems for microwave activation are either nonpolar or solvent-free systems, so the application of polar solvents should be always taken into account (Bogdal, 2005). In particular, for the reagents that are weak microwave absorbers, a polar solvent can initiate temperature rise of the reaction mixture—a small amount of a polar substance added to nonpolar medium can dramatically improve microwave absorption as well as temperature homogeneity of the sample (Table 2.3).

An interesting example of such a debate on the accelerating of chemical reactions by microwave irradiation is the ring opening polymerization of 2-phenyl-oxazoline recently reported by two research groups independently (Fig. 2.12) (Sinnwell et al., 2005; Wiesbrock et al., 2005).

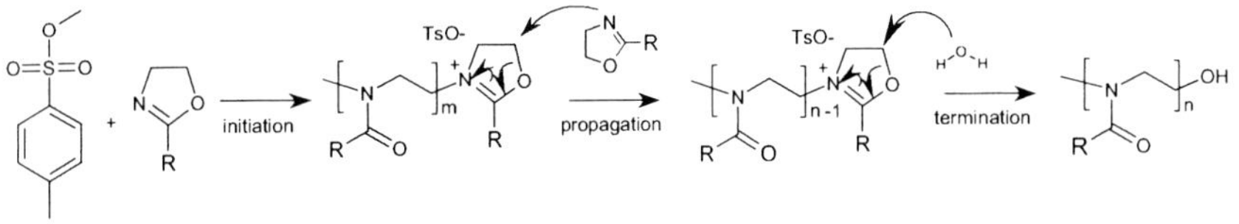

Fig. 2.12. The polymerization of 2-phenyl-oxazoline in the presence of methyltosylate as an initiator.

The polymerization of 2-phenyl-oxazoline in the presence of methyltosylate as a catalyst is an example of cationic ring opening polymerization, in which the charge is developed during the progress of the reaction and so, according to the previously mentioned concept of microwave irradiation, is supposed to accelerate the polymerization. In both reports, the polymerization reactions were carried out in an acetonitrile solution, which in turn as a polar solvent strongly interacts with microwaves and is supposed to mask or limit the influence of microwaves. The experiments were run in a 10-mL sealed vial in which 2.5 mL and 1.0 mL was placed, respectively, and irradiated in a dedicated for microwave synthesis single-mode microwave reactor (Sinnwell et al., 2005; Wiesbrock et al., 2005). In the former case, a comparison with thermal heating experiments showed a great enhancement in the reaction rates while the living character of the polymerization was conserved. Interestingly, the reaction rate coefficient under conventional conditions was the same for the reactions in an open and closed vessels (i.e., 1.8×10^{-4} sec^{-1}), while for the microwave experiments, the reaction rate coefficient was different for the reaction in open and closed reaction vessels (i.e., 6.0×10^{-4} sec^{-1} and 7.2×10^{-4} sec^{-1}, respectively) (Sinnwell et al., 2005). In the next report, it was found that upon enhancing the reaction rate by factors of up to 400 going from 80 to 200 °C, activation energies for the polymerization (activation energy [E_A]: 73 to 84 KJ/mol) were within the range of values obtained with conventional heating; however, the values of activation energies for conventional heating experiments (E_A) were taken from the literature (Seagusa et al., 1972). The first-order kinetics of the monomer conversion and livingness of the polymerization were maintained. Moreover, the polymerization can be carried out in concentrated solutions or even bulk conditions to afford well-defined polymers (PDI <1.20) (Wiesbrock et al., 2005).

For the purpose of this book, the values of the reaction rate coefficients estimated from both reports are calculated at 125 °C and compiled in Table 2.7.

For both reports, the reaction rate coefficients have similar values for the reactions under microwave conditions, whereas for polymerization under conventional conditions, the value of the reaction rate coefficient is lower

Table 2.7. Reaction rate coefficients k (sec^{-1}) for the polymerization of 2-phenyl-2-oxazoline in an acetonitrile solution in the presence of TsOMe at 125 °C.

Reaction conditions	Sinnwell et al., 2005	Wiesbrock et al., 2005
MW	1.26×10^{-2}	1.25×10^{-2}
Δ(oil bath)	3.27×10^{-3}	—

than that under microwave conditions as was measured for the first report (Sinnwell et al., 2005). In the second case, the calculation of the value of the reaction rate coefficient under conventional heating conditions gives the coefficient in the same range as for the reaction under microwave conditions, which is expected because it was estimated with the similar values of E_A for both microwave and conventional conditions (Wiesbrock et al., 2005).

In our opinion, it is a pity that reaction rate coefficients and E_A values were not measured and calculated for conventional conditions in this report because it might give a bit deeper view into the activation of chemical reactions under microwave conditions in comparison with conventional heating conditions.

Solid State Polymerization

The enhancement of solid state polycondensation for poly(ethylene terephthalate) (PET) as well as Polyamide 66 was reported in a fluidized bed reactor (Mallon et al., 1998). The apparatus for the polymerization is shown in Figure 2.13. The thermocouple inside the microwave cavity was removed prior to the experiments to avoid arching, and actual temperature control was done by manipulating the inlet gas temperature. In the apparatus, the field strength was continuously varied by manipulating a length of polypropylene tubing with water inside. The more water inside the polypropylene tube, the greater was the rate of microwave absorption and the lower was the overall field strength inside the microwave cavity (Mallon et al., 1998).

For each polymerization, the reactor charged with PET or Polyamide 66 pellets (approximately 0.9 g) was purged with high-purity nitrogen for 2 minutes (flow rate was ≈3 L/min). In the case of the polymerization of PET, the temperature was increased in steps to precrystallize PET to avoid sticking at reaction temperature. After purging, the external gas heaters were turned on, the inlet temperature setpoint was set at 132 °C, and the flow rate was increased to 17 L/min. The first crystallization step was allotted to 8 minutes. Next, the temperature setpoint was raised to 190 °C for 10 min-

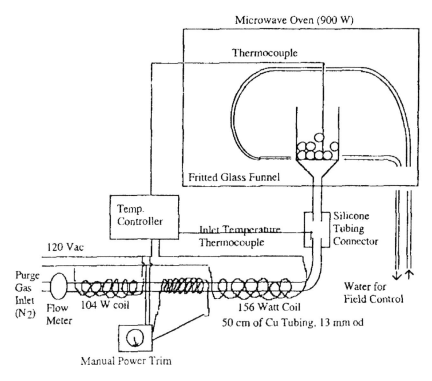

Fig. 2.13. The apparatus for the polycondensation of the PET and PA-66. Reprinted from Mallon, F.K., Ray, W.H. 1998. J. Appl. Polym. Sci. 69:1203, with permission.

utes. Finally, the temperature setpoint was raised to the reaction temperature. If this experiment involved microwaves, the field was switched on at this point and time from the zero time for the experiment was about 25 minutes. The samples were heated or irradiated for 6 to 7 hours; the same procedure was uses for all the experiments with PET and Polyamide 66 (Mallon et al., 1998).

Interestingly, it was found that the increase in polymerization rate was not due to an increase in polymerization temperature, but the effect was consistent with directed heating of the condensate leading to enhanced diffusion rates (Table 2.8). In the case of microwave PET polymerization, it was noticed that an increase in polymerization rate for different functional groups (i.e., the hydroxyl and carboxyl functions) was not monotonic, and it was higher for hydroxyl end groups. In general, the enhancement of solid-state reaction rate due to microwaves was about equivalent to an increase in reaction temperature of between 10 and 15 °C (Mallon et al., 1998).

Table 2.8.   Influence of microwave irradiation on PET polymerization in solid state.

Conditions	Diffusion coefficient at 220 °C Both ethylene glycol and water (cm²/sec) × 10⁶	Activation energy (cal/mol)
Δ	1.19	16.67
MW	3.55	12.19

Reprinted from Mallon, F.K., Ray, W.H. 1998. J. Appl. Polym. Sci. 69:1203, with permission.

Resin Transfer Molding

Microwave irradiation can be applied for resin transfer molding (RTM), and a specialized in-line microwave resin preheating system for such a purpose was developed (Fig. 2.14) (Hill et al., 1998; Johnson et al., 1998).

The system consists of a microwave 500- to 5000-W variable power generator operating in 2.46 GHz connected to the TM_{020} cylindrical applicator via a section of rectangular waveguide. Computer control enables remote operation of the microwave generator for a predefined heating period while simultaneously monitoring its efficiency via calorimetric measurements of the power dissipated with the water load. To demonstrate the performance of the cylindrical mode applicator, a stationary volume of resin was heated and the temperature distribution across the pipe section was measured via a circular array of 70 thermocouples (Hill et al., 1998; Johnson et al., 1998).

RTM has been limited to low-volume production due to protracted component cycle times. The principal cause of extended cycle time is the thermal quench near the injection gate. It occurs when cold resin enters the heated mold. The period required for the mold and resin to recover the lost heat and cure lengthens the cycle time. One method to decrease thermal quench and to reduce the cycle time is to preheat the resin prior to injection. For this purpose, a small volume of resin, having a low thermal inertia, passes through the in-line system, affording a rapid heating response to variations in the input of microwave power. The heating profile of the TM_{020} mode cylindrical applicator is analogous to the laminar flow profile, producing zero heat at the PTFE pipe wall where the resin is stagnant and maximum heat along the pipe axis where the velocity is highest (Fig. 2.15).

Using this system, the resin can be heated accurately to a constant, predefined temperature. Furthermore, prescribing an analytical heating func-

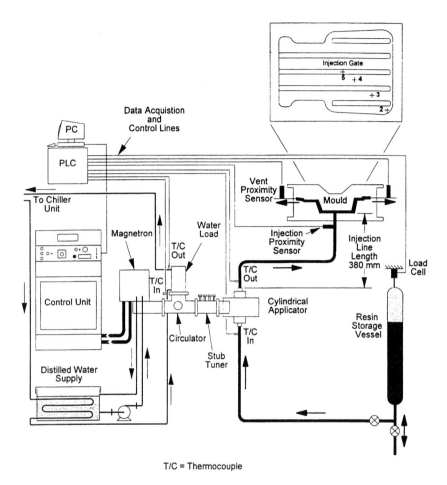

Fig. 2.14. Schematic of facility for microwave-assisted RTM. Reprinted from Johnson, M.S., Rudd, C.D., Hill, D.J. 1998. Composites Part A 29A:71, with permission.

tion allows profiling of the resin temperature during the injection phase. The cure sequence can be controlled using a ramping profile, with the additional benefit of a lower resin viscosity for improved flow through the mold. The molding was made with unsaturated polyester resin (Synolac 6345). As a result, impregnation and cycle times are reduced considerably. The increase of the resin temperature from 22 to 40 °C caused a 70% decrease in resin viscosity and reduced the impregnation time by 41% and the cycle time by 24% (mold temperature 60 °C) (Hill et al., 1998; Johnson et al., 1998).

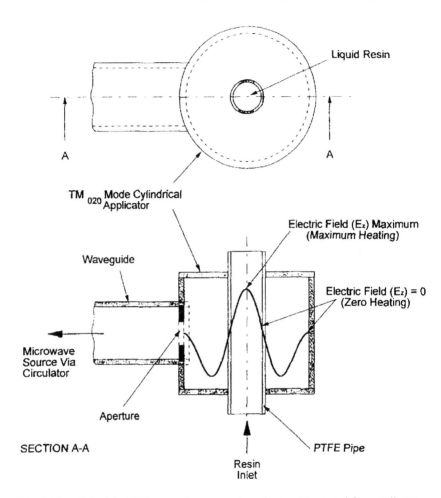

Fig. 2.15.   Cylindrical TM_{02n} mode resonant applicator. Reprinted from Hill, D.J., Rudd, C.D., Johnson, M.S. 1998. J. Microwave Power Electromag. Energ. 33:216, with permission.

Pulse Microwave Irradiation and Temperature Control

From the present chapter, it can be seen that there are many factors that have an influence on polymer synthesis and processing under microwave conditions. Not all materials are suitable for microwave applications, and a special characteristic of every process has to be matched. Therefore, a true cross-disciplinary approach has to be considered to fully understand all of the limitations and advantages of microwave processing. Improper application of microwave irradiation will usually lead to disappointments, while

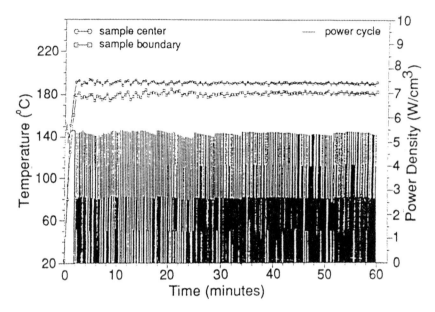

Fig. 2.16. Temperature–time profiles during controlled pulsed microwave curing of DER332/DDS resins. Reprinted from Jow, J., DeLong, J.D., Hawley, M.C. 1989. SAMPE Q. 20:46, with permission.

proper understanding and use of microwave power can bring greater benefits than expected. For example, three typical temperature–time stages can be observed during polymerization reactions:

* Initial temperature rise by direct heating of monomer(s) where the temperature rose slowly
* Significant temperature peak with maximum temperature due to the exothermic reaction
* Free convective cooling to an ambient temperature indicating an end to the exothermic reaction processes

Fast exothermic reaction heating usually accelerates the temperature rise and gradient inside the samples. Neither continuous microwave or thermal processing can be effectively controlled in order to maintain constant temperature–time profile through the entire process. However, pulsed microwave heating can be used to control temperature and eliminate the exothermic temperature peak, to maintain the same temperature at the end of reaction (Jow et al., 1989).

For example, the controlled pulse system has a unique feature of processing epoxy/amine resins at a higher constant cure temperature below thermal degradation temperature and without the exothermic temperature peak that always occurs in thermal or continuous microwave processing (Fig. 2.16) (Jow et al., 1989). Temperature profiles during the pulsed power curing of the epoxy/amine resins at 190 °C for 60 minutes are presented in Figure 2.16. It is shown that the temperature gradient inside the sample between its center and boundary developed as the setpoint temperature is increased. This is due to the greater boundary heat transfer at a higher cure temperature for microwave processing. Also, the controlled pulse microwave experiments at the temperature of 280 °C were successfully performed without thermal degradation of the cure samples. While an epoxy/amine sample was cured under conventional heating conditions at 240 °C for 60 minutes, thermal degradation of the samples was observed owing to the maximum exothermic peak above 300 °C during the cure process.

In summary, a survey of the past achievements in polymer synthesis and polymer composites can be found in review papers and books (Bogdal et al., 2003; Jacob et al., 1995; Mijovic et al., 1990; Parodi, 1999; Thostenson et al., 1999; Wei et al., 1996), and the fundamentals of electromagnetic heating and processing of polymers, resins, and related composites have also been summarized (Parodi, 1996).

References

Albert, P., Holderle, M., Mulhaupt, R., Janda, R. 1996. Acta Polym. 47:74.

Akyel, C., Bilgen, E. 1989. Energy 14:839.

Bogdal, D., Penczek, P., Pielichowski, J., Prociak, A. 2003. Adv. Polym. Sci. 163:193.

Bogdal, D., Gorczyk, J. 2004. J. Appl. Polym. Sci. 94:1969.

Bogdal, D. 2005. Microwave-Assisted Organic Synthesis, One Hundred Reaction Procedures. Elsevier. Amsterdam.

Bogdal, D., Matras, K. 2006a. Polymer chemistry under action of microwave irradiation. In: Microwaves in Organic Synthesis. Wiley-VCH. Weinheim.

Bogdal, D., Bednarz, S., Lukasiewicz, M. 2006b. Tetrahedron 62:9440.

Correa, R., Gonzalez, G., Dougar, V. 1998. Polymer 39:1471.

de la Hoz, A., Diaz-Ortiz, A., Moreno, A. 2005. Chem. Soc. Rev. 34:164.

Hill, D.J., Rudd, C.D., Johnson, M.S. 1998. J. Microwave Power Electromag. Energ. 33:216.

Jacob, J., Chia, L.H.L., Boey, F.Y.C. 1995. J. Mater. Sci. 30:5321.

Johnson, M.S., Rudd, C.D., Hill, D.J. 1998. Composites Part A 29A:71.

Jow, J., DeLong, J.D., Hawley, M.C. 1989. SAMPE Q. 20:46.

Krieger, B. 1992. Polym. Mater. Sci. Eng. 66:339.

Krieger, B. 1994. Mater. Res. Soc. Symp. Proc. 347:57.

Loupy, A. 2006. Microwaves in Organic Synthesis. Wiley-VCH, Weinheim.

Mallon, F.K., Ray, W.H. 1998. J. Appl. Polym. Sci. 69:1203.

Mijovic, J., Wijaya, J. 1990. Polym. Composites 11:191.

National Academy Press. 1994. Microwave Processing of Materials. National Academy Press.

Nishii, M. 1968. J. Osaka Dent. Univ. 2:23.

Nuchter, M., Muller, U., Ondruschka, B., Tied, A., Lautenschlanger, W. 2003. Chem. Eng. Technol. 26:1207.

Nuchter, M., Ondruschka, B., Bonrath, W. Gum, A. 2004. Green Chem. 6:128.

Palacios, J., Velverde, C. 1996. N. Polym. Mater. 5:93.

Parodi, F. 1996. Physics and chemistry of microwave processing. In: Comprehensive Polymer Science. 2nd Supplement. Aggarwal, S.L., Russo, E. (Eds.). Pergamon. Oxford:669.

Parodi, F. 1998. Chim. Ind. 80:55.

Parodi, F. 1999. Microwave heating and the acceleration of polymerization processes. In: Polymers and Liquid Crystals, Andrzej Wlochowicz (Ed.). Proc SPIE 4017:2.

Perreux, L., Loupy, A. 2001. Tetrahedron 57 :9199.

Perreux, L., Loupy, A. 2006. Nonthermal effects of microwaves in organic synthesis. In: Microwaves in Organic Synthesis, A. Loupy (Ed.). Wiley-VCH, Weinheim.

Schumb, W.C., Rittner, E.S. 1940. J. Am. Chem. Soc. 62:3416.

Seagusa, T., Ikeda, H. 1973. Macromolecules 6:808.

Seagusa, T., Ikeda, H., Fujii, H. 1972. Macromolecules 5:359.

Sinnwell, S., Ritter, H. 2005. Macromol. Rapid Commun. 26:160.

Thostenson, E.T., Chou, T.W. 1999. Composites Part A 30:1055.

Vanderhoff, J.W. 1969. US Patent 3 432 413.

Wei, J.B., Shidaker, T., Hawley, M.C. 1996. TRIP 4:18

Wiesbrock, F., Hoogenboom, R., Leenen, M.A.M., Meier, M.A.R. Schubert, U.S. 2005. Macromolecules 38: 5025.

Wiesbrock, F., Hoogenboom, R., Schubert, U.S. 2004. Macromol. Rapid. Commun. 25:1739.

Williams, N.H. 1968. J. Microwave Power 123:2.

Zhu, X., Chen, J., Zhou, N., Cheng, Z., Lu, J. 2003. Eur. Polym. J. 39:1187.

3

THERMOPLASTIC POLYMERS

Microwave irradiation has been used for the synthesis, processing, and joining as well as for the modification of a great number of thermoplastic polymers. During microwave irradiation of thermoplastic polymers, in particular semicrystalline thermoplastics, heating can be difficult until a so-called critical temperature is reached, at which point the dielectric loss factor can increase significantly (Chen et al., 1993). The critical temperature is related to the increase of molecular motion of polymer chains but is not necessarily the same as the glass transition temperature (T_g) of polymer. If the critical temperature is reached, rapid heating can occur (Zhong et al., 2003). Thus, the heating behavior during microwave irradiation of thermoplastic polymers is opposite from that of thermosetting resins (see Chapter 4), which in fact absorb less and less microwaves when they are being cured because of the formulation of cross-linked structures, increasing viscosity, and decreasing motion of polymer molecules. Their dielectric loss factor decreases significantly at the end of curing process.

Because in amorphous polymers the molecules are not restricted by the crystal lattice and they are more mobile than in semicrystalline polymers, amorphous polymers usually absorb more microwave energy and can be heated more effectively (Zhong et al., 2003)

Chain Polymerizations

Chain polymerization reactions under microwave irradiation were investigated for free radical polymerization and controlled "living" polymerization as well as ring-opening polymerization.

Free Radical Polymerization

Radical polymerizations of olefinic monomers are among the most important types of polymerization reactions. Microwave irradiation has been ap-

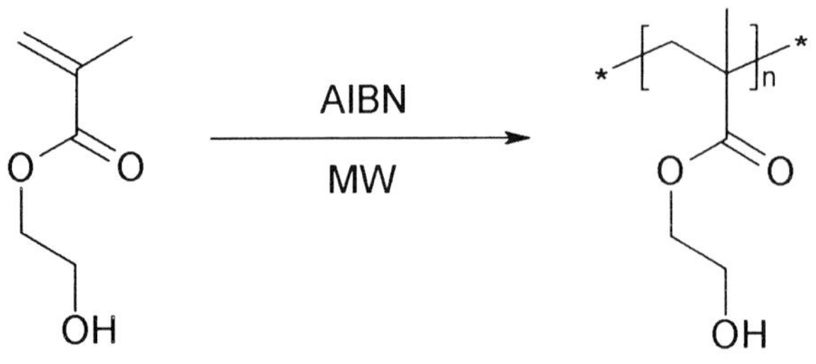

Fig. 3.1   Copolymerization of 2-hydroxy-ethyl methacrylate (HEMA) with methyl methacrylate (MMA).

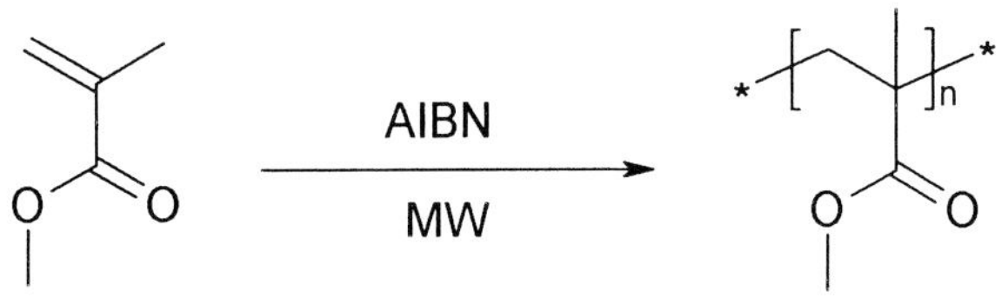

Fig. 3.2   Polymerization of methyl methacrylate (MMA).

plied to free radical polymerization and copolymerization reactions of various unsaturated monomers. For instance, the bulk polymerization of 2-hydroxyethyl methacrylate (HEMA) was investigated under microwave conditions because HEMA is a polar species that bears ester as well as alcohol functions capable of interacting and absorbing microwaves (Fig. 3.1) (Teffal et al., 1983). Therefore, the reactions were carried out without any radical initiator, and the liquid monomer polymerized to form a solid material that was insoluble with all the usual solvents but swelled in water. Thus, it was demonstrated that the radical polymerization can be achieved under microwave irradiation with neat reagents.

In a similar case, it was demonstrated that in the bulk copolymerization of HEMA with methylmethacrylate (MMA), microwave-assisted polymerizations gave copolymers with molecular weight twice as high and narrower molecular weight distribution in comparison to the copolymers obtained under conventional thermal conditions (Palacios et al., 1992).

The bulk polymerization of MMA (Fig. 3.2) was also investigated using a variable power of microwave irradiation, which led to the reaction enhancement by approximately 130% to 150% compared with the conventional methods (Fig. 3.3). Also, the limiting conversion of the reaction varied for the thermal and microwave polymerization (Chia et al., 1995).

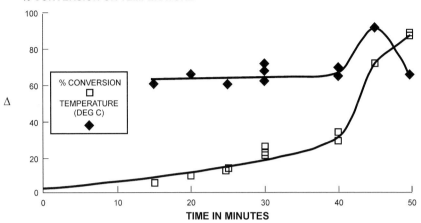

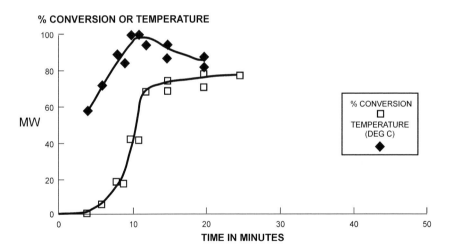

Fig. 3.3 Thermal profiles and conversions for conventional (Δ) and microwave (MW) polymerization of methyl methacrylate (MMA). Reprinted from Chia, L.H.L., Jacob, J., Boey, F.Y.C. 1995. J. Mat. Proc. Techn. 48:445.

The thermal polymerization of MMA at the comparative temperatures of 69, 78, and 88 °C displayed a limiting conversion of about 90%, while limiting conversion of microwave polymerization declined in following order: 200 W, 88% > 300 W, 84% > 500 W, 78%. The reactions were run in a sample vial with a narrow tube of 10 mm diameter and 2 mL capacity, in which 4 mg of azobisisobutyronitrile (AIBN) was placed with 0.5 mL of MMA. The microwave polymerization was conducted in a multimode cavity, which was designed with a rotating platform to prevent formation of "hot spots" due to nonuniform heating. The NMR analysis showed that the tacticities of the polymers for thermal and microwave polymerization are similar, suggesting that the polymerization process might not follow a different mechanism under microwave conditions (Chia et al., 1995).

The polymerization of methylacrylate (MA) with AIBN as an initiator under microwave irradiation was carried out under similar conditions (Jacob et al., 1997). To prepare samples for polymerization, 4.1 mg of AIBN (0.85 wt%) was taken in a 4 mL sample vial of 15 mm diameter with 0.5 mL (478 mg) of MA. The reaction rate enhancement of microwave polymerization compared with the thermal method was found to be as follows: 500 W, 275%; 300 W, 220%; and 200 W, 138%. Even though the comparable temperature at variable power was the same, 52 °C, the reaction rate enhancement increased with the increase in microwave power indicating a significant correlation between the reaction rate enhancements and the level of microwave power used (Jacob et al., 1997).

In turn, the bulk polymerization of styrene at two different power levels (300 and 500 W) was conducted in a multimode microwave cavity (Chia et al., 1996). The reactions were run in 2-mL sample vials of 10 mm diameter, in which 23.0 mg of AIBN was placed with 455 mg of styrene (Fig. 3.4).

The conversion profiles of the microwave polymerization were significantly different from that of the thermal cure at the same temperature of 80 °C. The thermal cure was characterized by a gradual gel effect from 30% to 50% conversion, while the microwave cure at 300 W and 500 W was characterized by a sharp and large gel effect from 20% to 69% and from 20% to 65%, respectively (Fig. 3.5). Moreover, the comparison of thermal and microwave polymerization under similar conditions showed a reaction rate enhancement of 190% for 500 W and 120% for 300 W. Similar to the microwave polymerization of MMA (Chia et al., 1995), the limiting conversion of styrene decreased from 72% for conventional thermal condition to 69% at 300 W and 65% at 500 W of microwave irradiation power. Finally, it was stated that comparison of kinetic results of microwave-induced reactions should con-

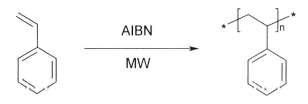

Fig. 3.4   Polymerization of methyl methacrylate (MMA).

sider the temperature as well as the power of microwave irradiation due to different energy supplied to the reaction system (Chia et al., 1996).

Later, the bulk polymerization of various vinyl monomers (i.e., vinyl acetate, styrene, MMA, and acrylonitrile) was studied (Porto et al., 2002). The

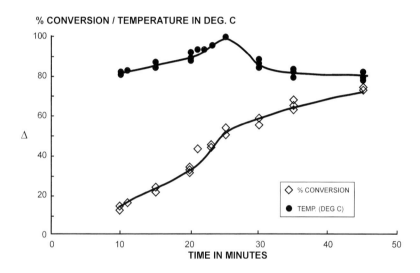

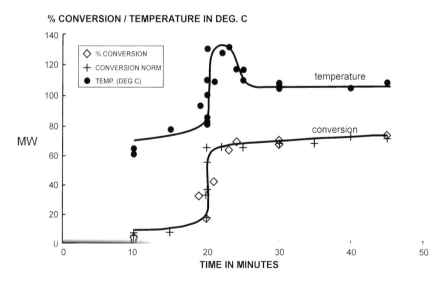

Fig. 3.5. Thermal profiles and conversions for conventional (Δ) and microwave (MW) polymerization of styrene. Reprinted from Chia, H.L., Jacob, J., Boey, F.Y.C. 1996. J. Polym. Sci. Part A Polym. Chem. 34:2087.

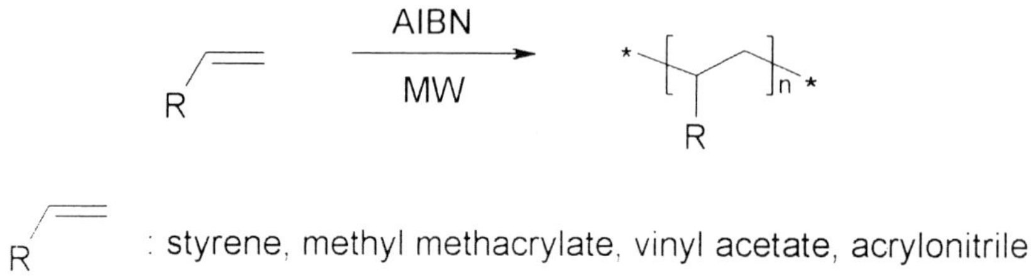

R ⟋═   : styrene, methyl methacrylate, vinyl acetate, acrylonitrile

Fig. 3.6   Polymerization of various vinyl monomers (i.e., vinyl acetate, styrene, methyl methacrylate, and acrylonitrile).

reactions were carried out in the presence of catalytic amount of AIBN (or benzoyl peroxide). It was found that the rate of polymerization depends on the structure of monomers and power and time of microwave irradiation. In a typical experiment, 10.0 mL of each monomer along with 50 mg of AIBN was irradiated in a domestic microwave oven for 1 to 20 minutes to afford polymers polystyrene, poly(vinyl acetate), and poly(methyl methacrylate) (PMMA), with weight average molecular weights of 48,400; 150,200; and 176,700 g/mol, respectively (Fig. 3.6). In fact, the experiments were run without any temperature control.

In turn, free radical homopolymerizations and copolymerizations of styrene were performed in toluene and N,N-dimethylformamide (DMF) as solvents using different initiators (i.e., *tert*-butyl perbenzoate [tBPB], dibenzoyl peroxide [DBPO], di-*tert*-peroxide [DtBP], dicumyl peroxide [DCP], and lauryl peroxide [LP]; Fig. 3.7) with and without microwave irradiation (Stange et al., 2006).

For microwave experiments, a single-mode microwave reactor equipped with online infrared temperature sensor and compressed air system for cooling was used. In a typical run, 0.435 mmol of initiator (i.e., 81 µL of tBPB, 0.132 µL of DBPO, 79 µL of DtBP, 0.117 µL of DCP, or 0.173 µL of LP) and 4.54 g (0.0435 mol) of styrene were dissolved in 5.2 mL of solvent under argon atmosphere in a 100-mL flask equipped with a reflux condenser and a stirring bar. The reaction mixture was irradiated in the microwave reactor or heated in an oil bath to 100 °C for 45 minutes.

For the copolymerization reactions, the same amounts of initiators and solvents were mixed together with 2.27 g (0.0217 mol) of styrene and 2.18 g (0.0217 mol) of MMA. Only the homopolymerization of styrene under microwave irradiation in DMF with DBPO showed the significantly enhanced styrene conversion to approximately 46%; however, it was lower than the conversion under conventional conditions (i.e., an oil bath), which reached approximately 50%. Other initiators resulted in no or only slight increase in styrene conversions under microwave irradiation that dropped into the

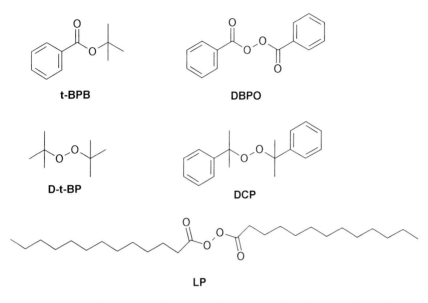

Fig 3.7 Chemical structures of initiators (i.e., *tert*-butyl perbenzoate [tBPB], dibenzoyl peroxide [DBPO], di-tert-peroxide [D-t-BP], dicumyl peroxide [DCP], and lauryl peroxide [LP]) used for radical polymerization.

range of 2% to 20%. In any case, DMF was required to gain increase in styrene conversion under microwave irradiation, which is obviously caused by the higher energy absorption by DMF compared with toluene in a given amount of time.

Significantly higher monomer conversions were observed under otherwise comparable conditions in the copolymerization of styrene and MMA. In this case, the monomer conversion of approximately 92% was observed under microwave irradiation in comparison to approximately 37% without microwave irradiation when tBPB was used as initiator in DMF. However, for the copolymerization with DBPO as an initiator, the conversions of 75% and 71% were obtained for the reactions under microwave and conventional conditions, respectively. Again, the use of toluene did not result in any enhancement by microwave irradiation.

Molar ratio of styrene and MMA in copolymers was determined by ^{1}H-NMR spectroscopy. For calculation of the molecular ratios of the copolymers, the aromatic protons of styrene were compared with the protons of the acrylic methoxy group. Only insignificant variations of styrene and MMA ratio in the copolymers compared with monomer feed were observed, regardless of reaction conditions. It was found that number average molecular weights of polystyrene samples and the copolymers were dependent on

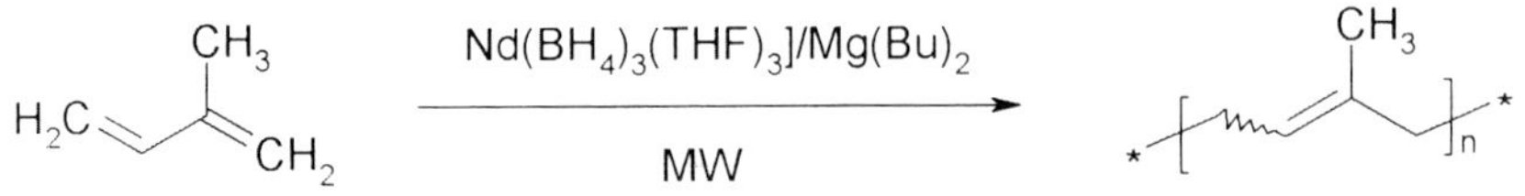

Fig. 3.8   Polymerization of isoprene in the presence of organolanthanide catalysts.

the initiator and dropped in the range of 0.76 to 19.90×10^4 g/mol and 1.30 to 36.90×10^4 g/mol for the homopolymerization and copolymerization experiments, respectively (Stange et al., 2006).

Recently, the polymerization of isoprene in the presence of organolanthanide catalysts under microwave irradiation was carried out (Zinck et al., 2005). The reactions were conducted in a single-mode microwave reactor with temperature infrared detection. The main power values necessary to reach and maintain temperature were 15, 32, 55, and 95 W for 60, 80, 100, and 120 °C, respectively (Fig. 3.8).

The polymerization of isoprene was carried out in a toluene solution, in which isoprene was mixed together with the catalyst [Nd(BH$_4$)$_3$(THF)$_3$] and a cocatalyst of either Mg(Bu)$_2$ and Al(Et)$_3$ in a 10-mL vessel that was sealed for microwave as well conventional experiments. The reaction time was in the range of 15 to 120 minutes; in some experiments, the reaction mixture was kept for 2 hours at room temperature prior to the reaction. The study showed an enhancement in reactivity under microwave condition in comparison with conventional conditions, while the selectivity was only slight modified. The highest yields (85% to 94%) of polyisoprene were obtained within 2 hours of reaction time at 80 °C to afford the polymer with number average molecular weight in the range of 17,000 to 27,000 g/mol and polydispersity index (PDI) of 1.6 to 2.5. Interestingly, the reaction at 120 °C afforded polyisoprene with a higher yield under conventional conditions, which was explained by a depolymerization reaction under microwave irradiation at high temperature (Table 3.1).

Chain transfer polymerization (telomerization) of poly-*N*-isopropylacrylamide (PNIPAM), poly-*N,N*-dimethylacrylamide (PNDMAM), and poly-*N*-[3-(dimethylamino)proply]acrylamide (PN3DMAPAM) as well as of copolymerization of PNIPAM and PNDMAM was also investigated under microwave irradiation (Fig. 3.9) (Fisher et al., 2005).

Under conventional conditions, the telomerization reactions were carried out in a superheated methanol solution (80 to 170 °C) in an autoclave in which 100 mmol of *N*-alkyl acrylamide was mixed together with 4 mmol of 3-mercaptopropionic acid, 0.5 mmol of AIBN, and 15 mL of methanol. For the microwave experiments, 50 mmol of *N*-alkyl acrylamide was mixed with

Table 3.1. Isoprene polymerization with [Nd(BH$_4$)$_3$(THF)$_3$]/Al(Et)$_3$.

Run	[Al]/[Nd]	Heating	Temperature, °C	Time, min	Yield, %	M$_n$, g/mol
24	5	Δ	80	480	45	11,400
25	20	Δ	80	480	72	8600
26	20	mv	80	480	20	7100
27	20	Δ	120	30	9	5300
28	20	mv	120	30	50	6500
29	20	mv	120	60	36	6100

Reprinted from Zinck, P., Barbier-Baudry, D., Loupy, A. 2005. Macromol. Rapid Commun. 26:46, with permission.

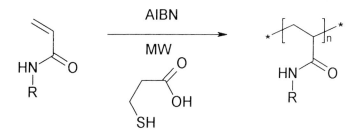

Fig. 3.9 Chain transfer polymerization (telomerization) of poly-*N*-isopropylacrylamide (PNIPAM), poly-*N,N*-dimethylacrylamide (PNDMAM), and poly-*N*-{3-(dimethylamino)proply}acrylamide (PN3DMAPAM).

2 mmol of 3-mercaptopropionic acid and 0.25 mmol of AIBN in an open flask that was irradiated in a domestic microwave oven. It was found that in superheated methanol, the reaction time was reduced by 66%, and moleculars weight and yield remained unchanged compared with standard reflux conditions. In microwave experiments, the reduction of the reaction time was even further; the average molecular weight dropped by 30% in these experiments (Table 3.2) (Fisher et al., 2005). However, it is hard to compare the bulk polymerization under microwave conditions without any temperature control with the solution polymerization under pressurized and non-pressurized conditions.

A number of (meth)acrylamides were synthesized under microwave irradiation in good yields from (meth)acrylic acid and aliphatic and aromatic amines under solvent-free conditions (Goretzki et al., 2004). It was found that an addition of polymerization initiator (i.e., AIBN) to the reaction mix-

Table 3.2. Comparison of the telomerization of the investigated monomers: (a) under standard conditions (methanol reflux, 65 °C, ambient pressure); (b) in superheated methanol; (c) under solvent-free conditions, by microwave induction.

Polymer	Reaction conditions	Reaction time	Monomer conversion, %	Average molar mass, g/mol
PNIPAM	65 °C (a)	3–5 h	84	2700
	80 °C (b)	1 h	57	2733
	120 °C (b)	1 h	82	2983
	350 W (c)	60 s	78	2447
	350 W (c)	150 s	81	1938
PNDMAM	80 °C (b)	1 h	70	2266
	120 °C (b)	1 h	92	2469
	350 W (c)	30 s	66	1677
	350 W (c)	150 s	73	2036
N3DMAPAM	80 °C (b)	1 h	71	—
	120 °C (b)	1 h	87	—
Co-PNIPAM/	80 °C (b)	1 h	98	2561
PNDMAM	120 °C (b)	1 h	98	2717
	350 W (c)	60 s	90	2080
	350 W (c)	120 s	75	2054

Reprinted from Fischer, F., Tabib, R., Freitag, R. 2005. Eur. Polym. J. 41:403, with permission.

ture led directly to poly(meth)acrylamides in a single-step reaction. For the polymerization protocols, 11.6 mmol of methacrylic acid was mixed with 11.6 mmol of an amine and 0.58 mmol of AIBN in a pressure-resistant test tube. The tubes were sealed and irradiated in a single-mode microwave reactor for 30 minutes at 140 W (Fig. 3.10). The experiments were run without a temperature control and comparison with conventional conditions.

Later, it was demonstrated that under microwave conditions it was possible to obtain chiral (R)-N-(1-phenyl-ethyl) methacrylamide directly from methacrylic acid and (R)-1-phenylethylamine under solvent-free conditions. An addition of free radical initiator AIBN led again in a single-step reaction to the formation of an optically active polymers that contained both methacrylamide and imide moieties (Fig. 3.11) (Iannelli et al., 2005; Iannelli and Ritter, 2005). In a typical polymerization reaction, 16.5 mmol of methacrylic acid was mixed together with 16.5 mmol of (R)-1-phenylethylamine and 0.83 mmol of AIBN in a pressure-resistant test tube. The tubes were sealed and irradiated in a single-mode microwave reactor for 30 minutes at 140 W at constant temperature 120 °C. It was found that microwave irradiation con-

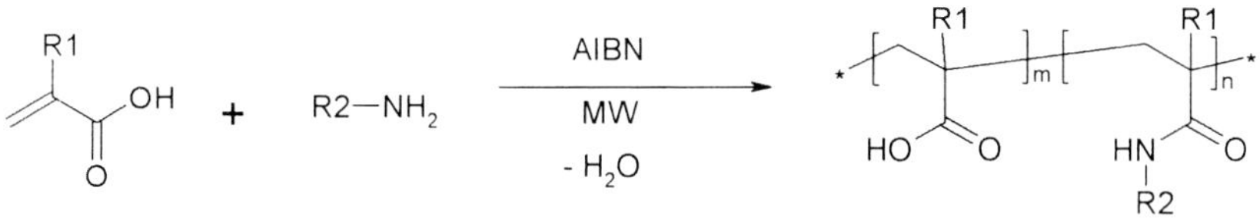

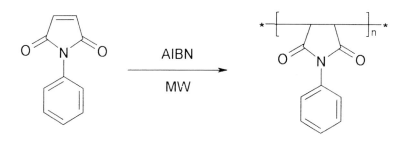

Fig. 3.10 Formation of poly(meth)acrylamides from the reaction of meth(acrylic) acid and various amines.

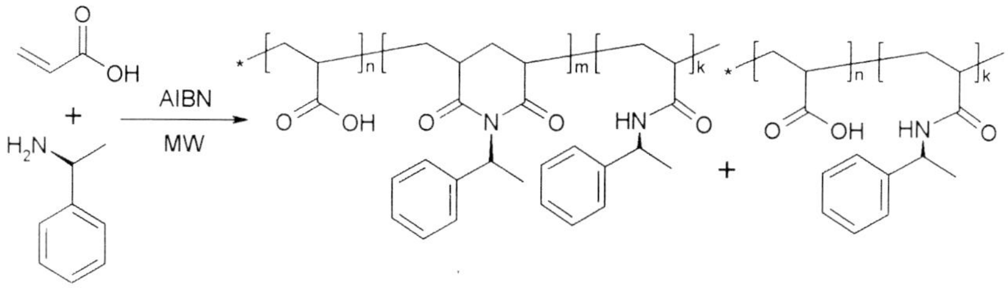

Fig. 3.11 Synthesis of chiral polymers from (*R*)-*N*-(1-phenyl-ethyl) methacrylamide and methacrylic acid and (*R*)-1-phenylethylamine.

Fig. 3.12 Polymerization of *N*-phenylmaleimide.

siderably accelerates the process of condensation between the acid and amine, which is also more selective under microwaves in comparison to thermal heating. The "one-pot" polymerization under microwave conditions afforded polymers with relatively high yields (80%), which depended on applied power. The yield under classic heating in an oil bath was only 40%.

In turn, the bulk polymerization of *N*-phenylmaleimide, which was prepared from maleic anhydride and aniline prior to the reaction, under microwave irradiation was presented (Fig. 3.12) (Bezdushna and Ritter, 2005).

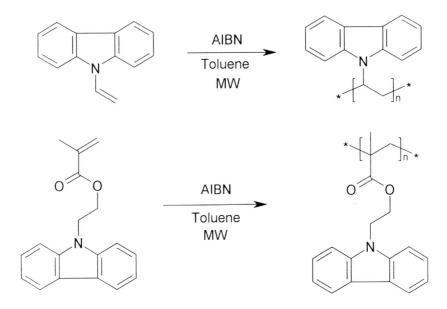

Fig. 3.13   Polymerization of carbazole-containing monomers [i.e., *N*-vinylcarbazole and 2-(9-carbazolylethyl)ethyl methacrylate].

For such a reason, 5.9 mmol of *N*-phenylmaleimide was mixed with 2.8 mmol of AIBN in a pressure-resistant test tube. The tube was sealed with a septum, flushed with nitrogen, and irradiated in a single-mode microwave reactor for 15 minutes at 90 °C (infrared pyrometry). The final yield of the polymer was 57%. The bulk homopolymerization under conventional conditions (an oil bath preheated to 95 °C) afforded the polymer with a relatively low yield of approximately 19%.

Free radical polymerization of carbazole containing monomers [i.e., *N*-vinylcarbazole and 2-(9-carbazolylethyl)ethyl methacrylate] under microwave irradiation was also carried out (Fig. 3.13) (Pajda et al., 2004). The reactions were carried out in pressure-resistant tubes in a solution by applying different solvents such as toluene, hexane, nitromethane, and diethylene glycol. In a standard protocol, 1 g of the monomer was mixed together with 1 mL of a solvent and 0.060 mmol of AIBN. The vessel was purged with argon, sealed, and irradiated in a single-mode microwave reactor for 10 minutes at 65 °C. After the precipitation, the polymers characterized by weight average molecular weights of 20,000 to 50,000 g/mol were afforded in high yields of 80% to 99%. Interestingly, in the experiments under conventional conditions (a preheated oil bath), the polymers were obtained in very low yields of approximately 1%.

The bulk polymerization of *N*-vinylcarbazole (NVC) in the presence of

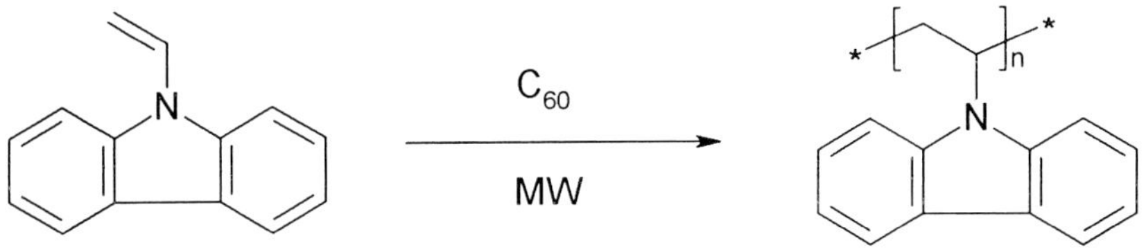

Fig. 3.14 Polymerization of *N*-vinylcarbazole (NVC) in the presence of fullerene-C_{60}.

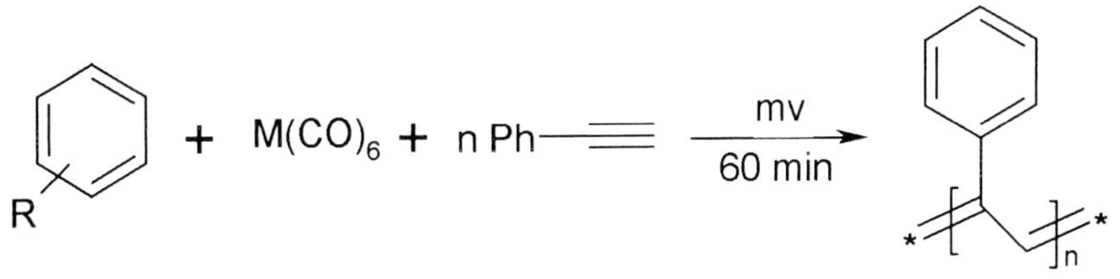

Fig. 3.15 Methathesis polymerization of phenylacetylenes in the presence in situ generated (arene)M(CO)$_3$ complexes.

fullerene-C_{60} as a charge-transfer initiator as well as initiator was studied (Chen et al., 2000). Microwave polymerizations were carried out in sealed argon-filled thick-walled Pyrex tubes filled with NVC (from 0.19 to 3.22 g) and C_{60} (0.02 g) and placed in a commercially available multimode microwave oven for 10 minutes. In comparison with polymerization under conventional conditions (water bath, 70 °C), microwave polymerization was found to be advantageous due to decrease in the reaction time and considerable improvement of yield of poly(*N*-vinylcarbazole) (PVK)—70% as opposed to 8% after 10 minutes of conventional heating (Fig. 3.14) (Chen et al., 2000).

In fact, during microwave experiments, temperature was not measured or even evaluated, so it is hard to make any yield comparison between these two techniques. For example, despite different heating methods, the molecular weights and polydispersities of the resultant polymers were similar.

Methathesis polymerization of phenylacetylenes under microwave irradiation was realized in the presence in situ generated (arene)M(CO)$_3$ complexes (Fig. 3.15) (Sundararajan et al., 1997). Microwave irradiation was carried out in a domestic microwave oven under a dry oxygen-free nitrogen atmosphere in a specially designed long-necked round-bottom flask that could withstand elevated pressure. The catalyst/monomer ratio was kept at a 1:50 level. For example, W(CO)$_6$ (0.199 mmol) was taken together with phenylacetylene (5.975 mmol) and toluene (0.5 mL) in a 1,2-dichloroethylene solution and irradiated for 5 minutes at time intervals of 10 minutes to cool down the reaction mixture. The reaction time was reduced to 1 hour in contrast to refluxing conditions of 24 hours. Among the group VIB metal carbonyls, $W(CO)_6$ was found to be the best catalyst precursor, while phenol showed the highest activity among arenes. Under such conditions, poly(phenylacetylene) was obtained in 67% to 75% yield.

The copolymerization reactions of maleic anhydride with allylthiourea (Lu et al., 1998) and dibenzyl maleate (Lu et al., 1997) were also studied. The reactions were carried out in solid state without any initiator in 10-mL vials, which were irradiated in a domestic microwave oven for 20 to 70 seconds. In the case of allylthiourea, the conversion of monomers reached approximately 20% after 70 seconds at 430 W, while yield of the copolymer obtained from dibenzyl malaete was 94% after 32 seconds of irradiation at 700 W. Under conventional conditions, the copolymer was produced with 89% yield after 2 hours at 45 °C (Lu et al., 1997). The copolymers were used as a metal-complexing agent (Lu et al., 1998) and suspension agent for the synthesis of superabsorbent oil resins (Lu et al., 1997).

More recently, the microwave copolymerization of dibutyltin maleate (DBTM) and allyl thiourea was investigated, under conditions specified above, to produce organo-tin copolymers (Lu et al., 2001). Under microwave irradiation, the copolymers could be synthesized with and without the presence of free radical initiator; however, in the presence of initiator, the conversion of monomers was slightly higher, reaching approximately 35%. Without using any initiator, the monomers did not copolymerize at all by conventional heating within 7 hours. The copolymers were applied to thermal stabilization of PVC. The thermal stability was increased from 180 °C for neat PVC to above 240 °C for PVC modified with the organo-tin copolymers (Lu et al., 2001).

The copolymerization of methacrylic acid, 2-(dimethylamino)ethyl ester, and allyl thiourea under microwave irradiation was carried out (Fig. 3.16) (Lu et al., 2004a). It was demonstrated that the copolymers can be used to coordinate Cu(II) to afford coordinated copolymers, which in turn can be applied a heterogeneous catalyst in the polymerization of MMA. The reactions were run in a modified domestic microwave oven with a continuous power regulation.

Controlled "Living" Radical Polymerization

Controlled "living" radical polymerization methods were developed for obtaining polymers with predetermined molecular weights, low PDI, specific functionalities, and diverse architecture compared with conventional free radical polymerization (Matyjaszewski et al., 2001).

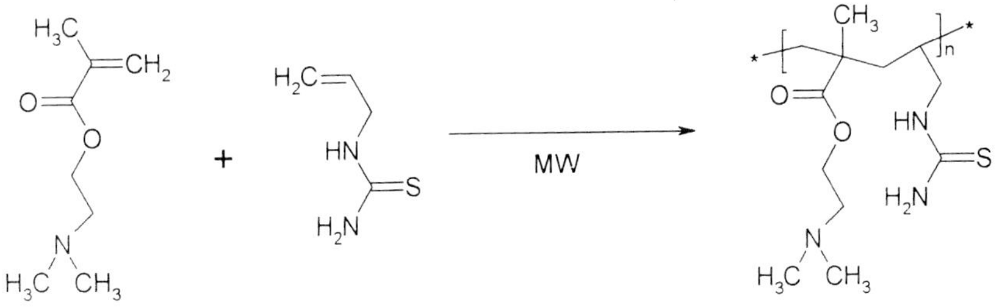

Fig. 3.16 Copolymerization reaction of methacrylic acid, 2-(dimethylamino) ethyl ester, and allthiourea.

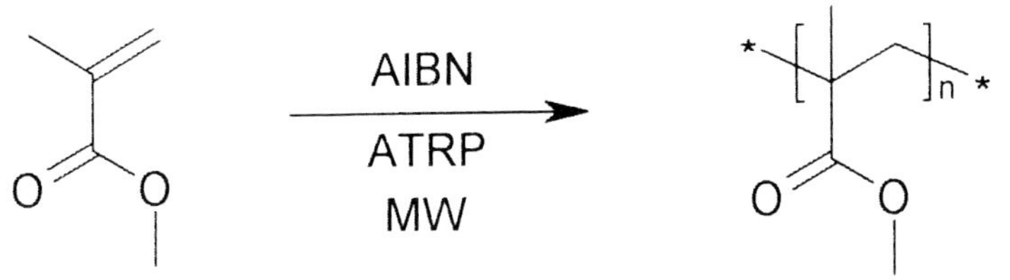

Fig. 3.17 Polymerization of methyl methacrylate (MMA) by atom transfer radical polymerization (ATRP).

The bulk polymerization of MMA by atom transfer radical polymerization (ATRP) under microwave irradiation was described in a number of reports (Fig. 3.17) (Chen et al., 2004; Lu et al., 2004b; Zhu et al., 2003).

The reactions were carried out in sealed tubes that were placed inside two-neck flasks filled with a solvent transparent for microwave (i.e., hexane, CCl_4). The flask was irradiated in a modified microwave oven so that temperature was controlled by the boiling point of a solvent in the flask. The reactions were run with different activator–initiator systems including benzyl chloride and bromide/$CuCl$/2,2′-bipyridine (Zhu et al., 2003), AIBN/$CuBr_2$/2,2′-bipyridine (Table 3.3) (Chen et al., 2004), and α,α′-dichloroxylene/$CuCl$/$N,N,N′,N″,N‴$-pentamethyldiethylenetriamine (Lu et al., 2004b).

It was found that in all the cases, microwave irradiation enhanced the rate of polymerization and gave polymers with narrower molecular weight distributions (PDI). Moreover, linear first-order rate plots, linear increase of the number average molecular weight with conversion, and low polydispersities were observed, which indicated that ATRP of MMA was controlled under microwave conditions (Fig. 3.18).

At the same time, the solution ATRP of MMA under microwave irradiation was also studied (Cheng et al., 2003a, 2003b). In a typical run of the poly-

Table 3.3. Data for conversion for ATRP of MMA under microwave irradiation at various ratios of $[MMA]_0/[AIBN]_0$.

$[MMA]_0/[AIBN]_0/$ $[CuCl]_0/[BPY]_0$	Time, min	Conversion, %	$M_{n, GPC}$	M_w/M_n	$M_{n, th}$
200:0.1:1:3	3	14.1	30,400	1.48	14,100
	5	19.1	30,800	1.50	19,100
	10	31.1	36,500	1.49	31,100
	15	38.0	43,900	1.61	38,000
	25	47.4	44,100	1.89	47,500
	35	49.3	44,800	1.82	49,300
200:0.05:1:3	5	16.5	30,600	1.42	33,000
	8	22.2	36,700	1.46	44,500
	10	24.3	40,600	1.61	48,600
	15	30.0	44,600	1.76	60,100
	30	33.9	42,600	1.57	67,800
200:0.03:1:3	5	9.9	42,300	1.35	33,100
	8	19.4	42,400	1.47	64,900
	10	21.3	53,500	1.51	71,200
	15	21.5	59,900	1.69	71,600

Condition: T = 69 °C; $[MMA]_0$ = 9.46 M; mv power 450 W; $M_{n, th} = [MMA]_0/[AIBN]_0 \times (1/2) \times M_{MMA}$.

Reprinted from Chen, G., Zhu, X., Cheng, Z., Xu, W., Lu, J. 2004. Radiat. Phys. Chem. 69:129, with permission.

merization, a small amount of CuCl/N,N,N',N'',N''-pentamethyldiethylenetriamine and ethyl 2-bromobutyrate (i.e., activator–initiator system) was placed in a 10-mL glass tube with 1.0 mL of DMF and 5.0 mL of MMA. The tube was sealed and placed in a two-neck reaction flask filled with hexane so that temperature was controlled by the boiling point of the solvent during reflux in a modified domestic microwave oven. Linear first-order rate plots, linear increase in the number average molecular weight with conversion, and low polydispersities were observed. It was found that microwave irradiation enhanced the rate of polymerization (Table 3.4). For example, after 2.5 hours of microwave irradiation, the monomer conversion reached 27%, and the polymers were afforded with number average molecular weight of 57,300 g/mol and a PDI of 1.19; while under conventional conditions, a similar conversion was achieved after 16 hours, and the polymers were characterized by number average molecular weight of 64,000 g/mol and PDI of 1.19 (Fig. 3.19). Similar results were obtained for the ATRP of *n*-octyl acrylate in an acetonitrile solution in the presence of 2-bromobutyrate, CuBr, and 2,2'-bipyridine under microwave conditions (Xu et al., 2003).

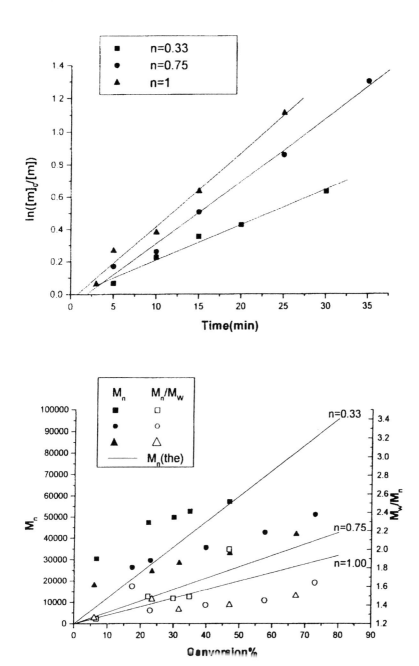

Fig. 3.18 Kinetics of atom transfer radical polymerization (ATRP) for methyl methacrylate (MMA) at different initiator concentration and dependency of M_n and M_w/M_n on conversion for ATRPT of MMA. Reprinted with the permission from Lu, X., Zhu, X., Cheng, Z., Xu, W., Chen, G. 2004b. J. Appl. Polym. Sci. 92:2189.

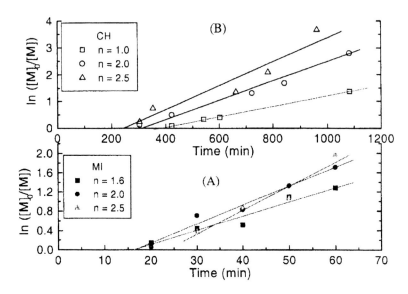

Fig. 3.19   Kinetic of solution of methyl methacrylate (MMA) in DMF under microwave (**A**) and conventional conditions (**B**) at different initiator concentration. Reprinted from Cheng, Z., Zhu, X., Zhang, L., Zhou, N., Xue, X. 2003a. Polym. Bull. 49:363.

Table 3.4.   Comparison of the propagation rate (kp^{app}) for the solution RATRP of MMA at different concentrations of initiator (MI—microwave, CH—conventional heating).

n	k_p^{app} (MI) (sec^{-1})	k_p^{app} (CH) (sec^{-1})	k_p^{app} (PMI) (sec^{-1})/ k_p^{app} (CH) (sec^{-1})
1	—	3.25×10^{-5}	—
1.6	48.6×10^{-5}	—	—
2	65.3×10^{-5}	5.47×10^{-5}	11.9
2.5	82.7×10^{-5}	6.93×10^{-5}	11.9

Reprinted from Cheng, Z., Zhu, X., Zhang, L., Zhou, N., Xue, X. 2003a. Polym. Bull. 49:363, with permission.

Unlike the previous investigation, it was reported that the ATRP of MMA in a *p*-xylene solution under microwave conditions did not give any rate enhancement in comparison with conventional conditions (Fig. 3.20) (Zhang and Schubert, 2004). In a typical experiment, a stock solution of 70 mmol of MMA, 15 mL/g of *p*-xylene, 0.47 mmol of ethyl 2-bromoisobutyrate, and *N*-hexyl-2-pyridylmethanimine was divided into six vials containing 0.078 mmol of CuBr. The vials were sealed, purged with argon, and irradiated in

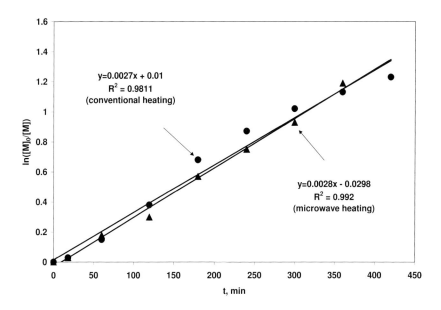

NHPM :

Fig. 3.20 Atom transfer radical polymerization (ATRP) of methyl methacrylate (MMA).

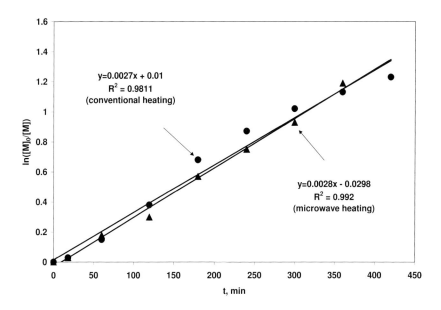

Fig. 3.21 Comparison of the kinetic plot of the atom transfer radical polymerization (ATRP) of methyl methacrylate (MMA) in DMF under conventional and microwave conditions. Reprinted from Zhang, H., Schubert, U.S., 2004. Macromol. Rapid Commun. 25:1225.

a single-mode microwave reactor for different periods up to 6 hours. The polymerization reaction exhibited good control in terms of linear first-order rate plots, linear increase of the number average molecular weight with conversion, and low polydispersities; however, they provided almost the same results as those performed under conventional conditions (Fig. 3.21).

Recently, the bulk ATRP polymerization under microwave irradiation was also used for the preparation of polyacrylonitrile (Hou et al., 2006). The polymerization of acrylonitrile was achieved by using $FeCl_2$, succinic acid as the catalyst, and 2-chloropropionitrile as the initiator taken in the molar ratio of 400:1:1:2. The reactions were performed in sealed tubes in a vacuum that were placed in a refluxing carbon tetrachloride solution irradiated in a modified household microwave oven. The $FeCl_2$/succinic acid ratio of 1:2 not only gives the best control of molecular weight and its distribution but also provides a rather rapid reaction rate under microwave conditions (apparent rate constant of $1.03 \times 10^{-4} \, sec^{-1}$), which was found to be higher than that under conventional heating (apparent rate constant of $2.2 \times 10^{-5} \, sec^{-1}$). In fact, under microwave irradiation, 30% monomer conversion was achieved to afford the polymer with number average molecular weight of approximately 90,000 g/mol and PDI of 2.35. Unfortunately, when $FeCl_2$ was replaced with CuCl, the ATRP polymerization of acrylonitrile did not show living characteristics.

In turn, bromo-double-terminated polystyrene (Br-PS-Br) and poly(methacrylate) (Br-PMMA-Br) with predesigned molecular weight and narrow polydispersity prepared by ATRP were reacted with excess amounts of fullerene-C_{60} in the presence of a CuBr/bipyridine (CuBr/bipy) catalyst system under microwave irradiation. As the result of the ATRA process, telechelic C_{60} end-capped polymers were obtained (Fig. 3.22) (Wu et al., 2006).

In a typical run, fullerene-C_{60}, CuBr/bipy, and bromo-terminated polymers were introduced into a two-neck round-bottom flask (250 mL) and de-oxygenated, and then toluene (40 mL) and DMF (10 mL) were added. The mixture was irradiated under nitrogen atmosphere in a modified domestic microwave oven for 20 minutes with constant power of 300 W. Under conventional conditions, the same flask was charged in a similar manner with the same amount of substrates and solvents and then under nitrogene in an oil bath for 8 hours at 110 °C and 90 °C for styrene and MMA, respectively (Table 3.5).

The results showed that microwave irradiation could significantly increase the rate of fullerenation reactions of bromo-end-capped polymers, while the physical properties and structure of the C_{60} end-capped polymers were not modified. However, it is hard to make strict comparisons because temperature was not determined during microwave experiments.

Solid-supported TEMPO-mediated controlled "living" polymerization, for the preparation of novel high-loading functionalized styrenyl resins, was reported by Wisnoski et al. (2003) (Fig. 3.23). The resin was prepared in a neat reaction of TEMPO-methyl resin with styrene derivatives. For example, 200 mg of TEMPO methyl resin suspended in 16.8 mmol of *p*-bromostyrene was transferred to a 5-mL vial, sealed, and placed in a single-mode microwave reactor. The mixture was irradiated for 10 minutes at 185 °C to af-

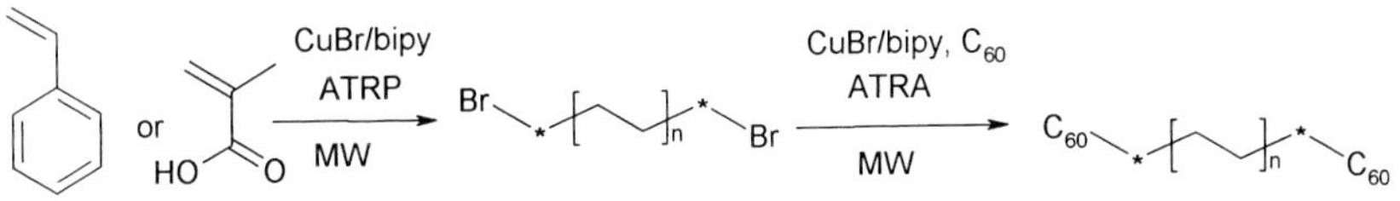

Fig. 3.22 Synthesis of telechelic C_{60} end-capped polymers.

Table 3.5. Results for the telechelic C_{60} end-capped products.

Product	Reaction time	M_n ($\times 10^4$)	M_w/M_n	C_{60} content, wt%	T_g, °C
C_{60}-PSt-C_{60}(MI)	20 min	1.129	1.150	10.03* 10.25† (12.75)	97.7
C_{60}-PMMA-C_{60}(MI)	15 min	2.511	1.532	6.35* 5.51† (5.73)	124.4
C_{60}-PSt-C_{60}(CH)	8 h	1.195	1.311	11.55* 12.41† (12.05)	98.0
C_{60}-PMMA-C_{60}(CH)	8 h	2.457	1.502	4.88* 5.41† (5.86)	123.9

*Calculated on the basis of the UV absorbance at 330 nm.

†Measured by TGA on the parent polymers and the C_{60} end-capped polymers; the data in parentheses are the theoretical value based on Mn and telechelic C_{60}-monoadduct structure.

Reprinted from Wu, H., Li, F., Lin, Y., Yang, M., Chen, W., Cai, R. 2006. J. Appl. Polym. Sci. 99:828, with permission.

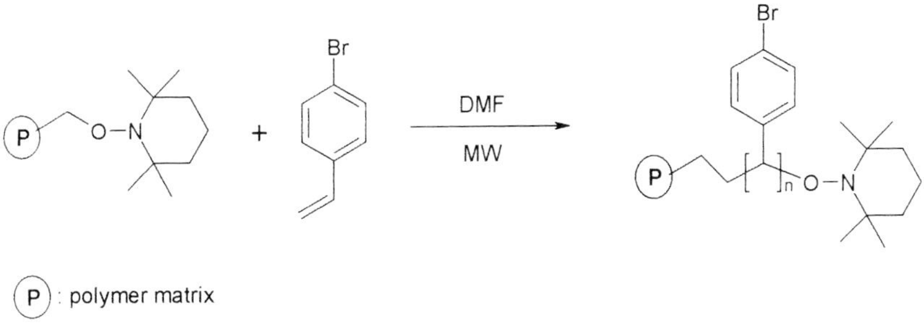

P : polymer matrix

Fig. 3.23 Solid-supported TEMPO-mediated controlled "living" polymerization of styrene monomers.

ford resin with 7.23-fold increase in mass. In applying this protocol, it was possible to obtain high-loading Rasta Merrifield resin (5.8 mmol/g). It is stressed that the microwave procedure was 150-fold faster compared with those described in the literature under conventional conditions.

Recently, TEMPO-mediated radical polymerizations of styrene in bulk were successfully performed under microwave irradiation (Li et al., 2006). The re-

Table 3.6.   Experimental results of thermal-initiated polymerization of styrene under microwave and conventional heating at 125 °C.

Method	Power (W)	Time (h)	Conversion (%)	M_n	M_w/M_n
Δ	—	3	0	—	—
Δ*	—	3	16	4400	1.25
Δ	—	6	9	4650	1.18
Δ	—	9	18	8100	1.23
MW	100	3	3	2570	1.14
MW*	100	3	41	9700	1.26
MW	200	2	8	4600	1.15
MW*	200	2	8	13,600	1.38
MW	200	3	37	11,100	1.38

*Conditions: $[styrene]_0 = 8.7$ M; $[BPO]_0 = 0.0305$ M; $[OH\text{-}TEMPO]_0 = 0.0366$ M. Others without BPO.

Reprinted from Li, J., Zhu, X., Zhu, J., Cheng, Z. 2006. Radiat. Phys. Chem. 75:253 with permission.

actions were carried out in a single-mode microwave reactor with temperature and power controlled by means of an infrared sensor (a pyrometer). In a typical polymerization procedure, styrene (2 mL, 17.4 mmol) was added to a dry ampule tube containing the solid mixture of benzyl peroxide (14.7 mg, 0.0609 mmol) and OH-TEMPO (12.6 mg, 0.0731 mmol). The contents were purged with nitrogen to eliminate oxygen for approximately 10 minutes. Then the ampule was flame-sealed, placed in a 10-mL pressure reaction vessel in the cavity of a single-mode microwave reactor, and heated by microwave irradiation at constant power in conjunction with adjusting cooling gas flow to the desired temperature to polymerize. In the case of conventional heating, the ampule was placed in an oil bath held by a thermostat at the desired temperature for polymerization. After desired polymerization time, the ampule was quenched in cold water to stop the polymerization and opened. The contents were dissolved in THF and precipitated into a large amount of methanol. The polymerizations were well controlled in terms of linear kinetic plots, linear increase in M_n of the polymers with increasing conversion and narrow PDI (1.14 to 1.38) was obtained (Table 3.6).

The power of microwave irradiation had considerable effect on the rate of the polymerization. The polymerization rates at appropriate power of microwave irradiation were faster than those under conventional heating conditions at the same polymerization temperature with or without benzyl peroxide (Table 3.7). Furthermore, it was proved by successful chain extension polymerization and NMR spectrum analysis that the nitroxide moiety did exist in the end of polymeric chain (Li et al., 2006).

Table 3.7. Experimental results of styrene polymerization at different temperatures under microwave irradiation.

Temperature (°C)	Conversion (%)	M_n	M_w/M_n
115	8	3250	1.22
120	18	5200	1.23
125	41	9700	1.26
130	66	14,300	1.31
135	76	15,000	1.41

Microwave power = 100 W; time = 3 h.
Reprinted from Li, J., Zhu, X., Zhu, J., Cheng, Z. 2006. Radiat. Phys. Chem. 75:253, with permission.

Also recently, nitroxide-mediated polymerizations of methyl- and *tert*-butyl acrylate in the presence of 2-methyl-2-(*N-tert*-butyl-*N*-(1(-diethylphosphono-2',2'-dimethylpropyl)aminoxyl)propanoic acid (MAMA) and the radical *N-tert*-butyl-*N*-(1-diethylphosphono-2,2-dimethylpropyl)nitroxide (SG1) as initiators were performed under microwave irradiation (Fig. 3.24) (Leenen et al., 2005).

The microwave-assisted polymerizations were performed in a single-mode microwave reactor equipped with a calibrated infrared sensor for the measurement of temperature. Vials filled with stock solutions containing the monomer, solvent, initiator, and free radical (if applicable) were irradiated. A typical stock solution contained approximately 50 wt% of the monomer and was composed of either 4.133 g of MA and 0.366 g of MAMA in 3.677 mL of dioxane (6 mol/L solution of the monomer) or 12.817 g of *tert*-butyl acrylate, 0.763 g of MAMA (and 29.4 mg of SG1, if applicable) in 10.599 mL of dioxane (4 mol/L solution of the monomer). Then 1 mL of the corresponding stock solution was placed into the vials. These were equipped with a stirring bar, capped, and bubbled for 10 minutes with argon to exclude oxygen. The thus-prepared vials were subsequently either irradiated in the microwave reactor or heated in an oil bath at 120 °C. In the latter case, the temperature of the oil bath had to be set to 133 °C in order to ensure an *internal* temperature of 120 °C in the solution in the (thick-walled) vials (as determined by reference measurements in DMF in open vessels). For the polymerization of *tert*-butyl acrylate, the monomer consumption followed first-order kinetics, while that of MA could be described with a kinetic model that includes the persistent radical effect.

Poly(methylacrylate)s with narrow molecular weight distributions (PDIs < 1.3) were obtained within 3 hours from the polymerization solutions. The control over the reaction could be preserved for monomer conversions of

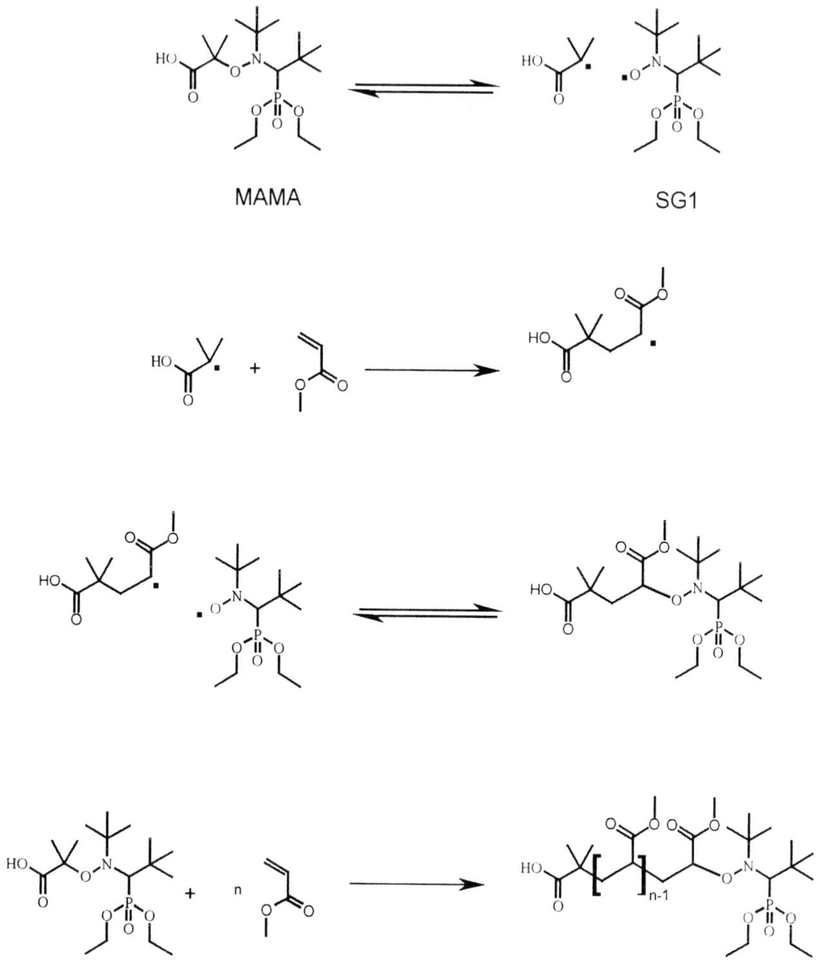

Fig. 3.24   Nitroxide-mediated polymerizations of methyl and *tert*-butyl acrylate.

up to 90%. Comparison experiments using an oil bath as heating source, however, showed a limited reproducibility and failed to yield polymers with similarly narrow molecular weight distributions (for high conversions). This observation was refereed to the superiority of the uniform, noncontact, and internal heating mode of microwave irradiation; any nonthermal microwave effects (like an acceleration of the polymerization) were not observed. For the polymerization of *tert*-butyl acrylate (which was performed exclusively under microwave irradiation), in contrast to MA, an excess of the nitroxide radical (SG1) had to be added to the polymerization solution in order to obtain polymers with the targeted narrowness of the molecular

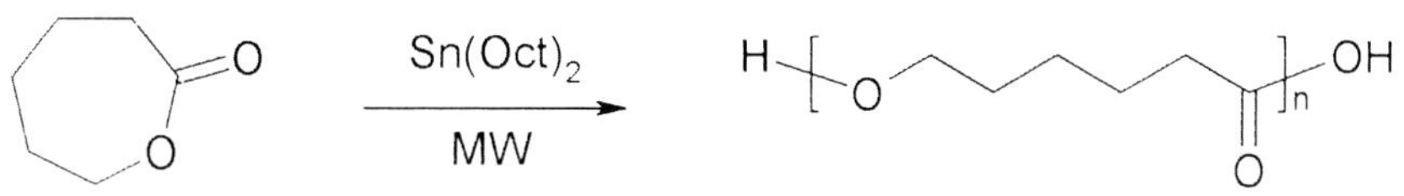

Fig. 3.25 Ring-opening polymerization of ε-caprolactone in the presence of Sn(Oct)₂ and zinc powder as catalysts.

weight distributions (PDI < 1.3). Thus, the reaction was decelerated, but monomer conversions of around 90% were achieved within 5 hours (Leenen et al., 2005).

Ring-Opening Polymerization

The ring-opening polymerization of ε-caprolactone under microwave irradiation was carried out in the presence of Sn(Oct)₂ and zinc powder as catalysts (Fig. 3.25) (Liao et al., 2002).

Typically, the reaction mixture that consisted of ε-caprolactone and catalyst in a vacuum sealed ampoule was irradiated in multimode microwave oven at different temperatures ranging from 80 to 210 °C. For example, poly(ε-caprolactone) with weight average molecular weight of 124,000 g/mol and yield of 90% was obtained after 30 minutes of irradiation at 680 W using 0.1% mol/mol Sn(Oct)₂, whereas the polymerization catalyzed by zinc powder afforded poly(ε-caprolactone) with weight average molecular weight of 92,300 g/mol after 30 minutes of irradiation at 680 W using 1% mol/mol zinc powder. Without microwave irradiation, the polymerization rate was considerably slower, at 120 °C poly(ε-caprolactone) was afforded with weight average molecular weight of 60,000 g/mol with Sn(Oct)₂ after 24 hours and 27,000 g/mol with zinc powder after 48 hours (Fig. 3.25).

Similar protocol was applied for the metal-free synthesis of poly(ε-caprolactone) from ε-caprolactone in the presence of benzoic acid (Yu et al., 2004). The molar ratios of ε-caprolactone to benzoic acid were in the range of 5 to 25, and the reaction mixture was heated at a so-called self-regulated temperature range of 204 to 240 °C in a multimode microwave reactor. The advantage of microwave protocol is an enhancement of propagation rate; however, above 240 °C degradation of poly(ε-caprolactone) became significant. With the metal–free method, weight average molecular weight was approximately 40,000 g/mol. The results of the polymerization reactions including number average molecular weights are presented in Table 3.8.

The ring-opening polymerization of DL-lactide in the presence of Sn(Oct)₂ under microwave irradiation was described (Fig 3.26) by Zhang

Table 3.8. Results of ring-opening polymerization of ε-caprolactone (monomer/initiator ratio of 5).

Heating	Reaction time, min	M_n, g/mol	M_n^{mv}/M_n^t	Monomer conversion, wt %
Δ	30	1500	6.6	92.1
mv		9900		92.0
Δ	60	1700	6.5	99.0
mv		11,100		95.0
Δ	100	2300	6.4	99.0
mv		14,700		94.9

Δ, thermal polymerization at 210 °C; mv, microwave-assisted polymerization at 680 W.
Reprinted from Yu, Z.J., Liu, L.J. 2004. Eur. Polym. J. 40:2213, with permission.

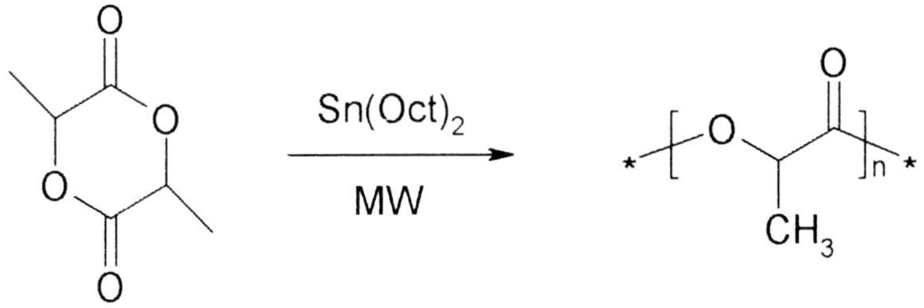

Fig. 3.26   Ring-opening polymerization of DL-lactide in the presence of Sn(Oct)$_2$.

and others (2004). It was stated that the polymerization of DL-lactide proceeded quickly, but no comparison to a conventional procedure was made. Under optimal conditions, poly(DL-lactide) (weight average molecular weight: 400,000 g/mol) was obtained with 90% yield after 10 minutes; however, it was stated that the degradation of the product was greatly influenced by microwave irradiation.

The polymerization of DL-lactide under microwave irradiation in an open vessel was performed as well (Jing et al., 2006). A mixture of DL-lactide and Sn(Oct)$_2$ in an open beaker was irradiated at a household microwave power of 450 W for some period of time, with a kind of assisted heating medium being used or not (Table 3.9). Also in a microwave oven, a certain amount of heating medium was placed in a heat-resistant container, and then the beaker containing reactant was placed on the heating medium during the course of microwave irradiation.

Table 3.9. Effect of heating medium on PDLLA.

Medium	Time (min)	$M_n \times 10^{-4}$	Color
Activated carbon	10	1.23	Brown
SiC	18	2.46	Pale yellow
Al$_2$O$_3$	60	2.30	Pale yellow
No media	65	2.22	Pale yellow

Conditions: Sn(Oct)$_2$/DLLA = 0.15 mol%; DLLA purity = 94.6 %.
Reprinted from Jing, S., Peng, W., Tong, Z., Baoxiu, Z. 2006. J. Appl. Polym. Sci. 100:2244, with permission.

To determine the structure and the yield of poly(DL-lactide), the crude polymer was dissolved in chloroform and precipitated in methanol. It was found that under the appropriate conditions such as carborundum (SiC) as heating medium, 0.15% catalyst, lactide with purity above 99.9%, 450 W microwave power, 30-minute irradiation time, and atmosphere, poly(DL-lactide) with a viscosity average molecular weight over 2.0×10^5 g/mol and a yield over 85% was obtained (Jing et al., 2006).

Recently, the ring-opening polymerization of ε-caprolactone in the presence of lanthanide halides as catalysts under microwave irradiation was also described (Barbier-Baudry et al., 2003). Opposite to the previous report, the reactions were run in open vessels in a single-mode microwave reactor. For such a purpose, 1 mL of ε-caprolactone was mixed with 2 mg to 50 mg of a catalyst and irradiated in a Pyrex tube at 200 to 230 °C. Highest number average molecular weight of polymers were obtained when the mixture of the monomer and catalyst was intensively heated in a microwave reactor so that the boiling point of ε-caprolactone was reached at approximately 1 minute. Number average molecular weights of the obtained polymers were between 2900 and 14,100 g/mol (Table 3.10). Compared to conventional thermal processes, under microwave conditions the polymers gained higher molecular weight and lower PDI. Moreover, in order to have the reactions entirely completed, longer thermal polymerization times were necessary.

More recently, oxazoline derivatives became the next group of cyclic monomers that was studied for ring-opening polymerization reactions under microwave irradiation (Fig. 3.27). For instance, 7.1 mL of 2-phenyl-2-oxazoline was mixed with 7.1 mL of acetonitrile together with 0.82 mmol of methyltosylate. Then 2.5 mL of this mixture was transferred to a 10-mL vial sealed with septum and irradiated in a single-mode microwave reactor for 30 to 150 minutes at 125 °C, which was monitored by a fiberoptic sensor. A comparison with thermal heating experiments showed a great en-

Table 3.10.   Results of ε-caprolactone polymerization.

Heating	Catalyst (mg)	Temperature, °C	Time, min	M_n, g/mol	M_w/M_n
Thermal	SmBr$_3$, 6H$_2$O (15)	200	30	3600	1.94
Thermal	SmCl$_3$, 6H$_2$O (11)	200	15	2300	2.00
mv	SmCl$_3$, 3THF (10)	200	45	14,100	2.20
mv	SmCl$_3$, 3THF (10)	230	3	13,500	2.91
mv	YbCl$_3$, 3THF (15)	230	3	11,300	2.57
mv	YbCl$_3$, 3THF (10)	200	15	11,800	1.82

Reprinted from Barbier-Baudry, D., Brachais, L., Cretu, A., Gattin, R., Loupy, A., Stuerga, D. 2003. Environ. Chem. Lett. 1:19, with permission.

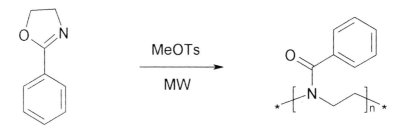

Fig. 3.27   Ring-opening polymerization of oxazoline derivatives.

hancement in the reaction rates while living character of the polymerization was conserved. For example, under microwave irradiation after 90 minutes, the conversion of the monomer was nearly quantitative (i.e., 98%). In contrast, the polymerization under conventional conditions showed only 71% conversion after 90 minutes. Interestingly, the reaction rate coefficient under conventional conditions was the same for the reactions in open and closed vessels (i.e., 1.1×10^{-2} min^{-1}), while for the microwave experiments, the reaction rate coefficient was different for the reaction in open and closed reaction vessels (i.e., 3.6×10^{-2} min^{-1} and 4.2×10^{-2} min^{-1}, respectively) (Sinnwell et al., 2005).

Then, the ring-opening polymerization reactions of a number 2-substituted-2-oxazolines (i.e., 2-methyl, 2-ethyl, 2-nonyl, and 2-phenyl) in the presence of methyltosylate as a catalyst were described (Fig. 3.27) (Hoogenboom et al., 2005; Wiesbrock et al., 2004, 2005a). The reactions were performed at the temperature range from 80 °C to 200 °C in a single-mode microwave reactor. In a typical run, 25 mL of stock solutions of monomer/initiator/

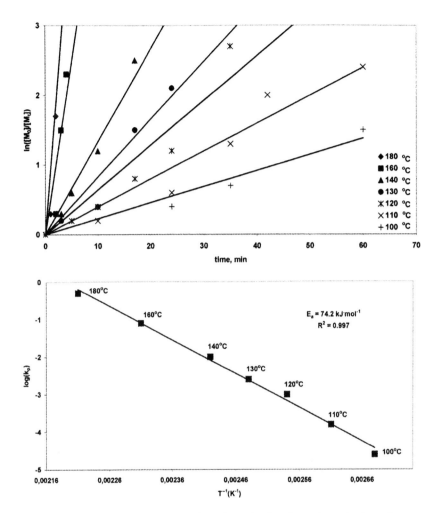

Fig. 3.28 Monomer conversion against time plot for the polymerization of 2-nonyl-2-oxazolie at different temperatures. Reprinted from Hoogenboom, R., Wiesbrock, F., Leenen, M.A.M., Meier, M.A.R., Schubert, U.S. 2005. J. Comb. Chem. 7:10.

solvent were prepared prior to the polymerization. These stock solutions were divided over the different reaction vials so that each experiment was performed in a 1-mL scale. It was found that upon enhancing the reaction rate by factors of up to 400 going from 80 to 200 °C (Fig. 3.28), activation energies for the polymerization (E_A = 73 to 84 KJ/mol) were within the range of values obtained with conventional heating (Table 3.11). The first-order kinetics of the monomer conversion and livingness of the polymer-

Table 3.11.   Activation energies EA and frequency factors A for the polymerization of 2-methyl-, 2-ethyl-, 2-phenyl-, and 2-nonyl-2-oxazoline.

Monomer	Frequency factor A, 10^8 L $\times$ mol^{-1} $\cdot$ s^{-1}	Activation energy E_A, kJ $\times$ mol^{-1}
2-Methyl-2-oxazoline	5.00 ± 1.20	75.4 ± 0.5
2-Ethyl-2-oxazoline	1.99 ± 0.85	73.4 ± 0.5
2-Phenyl-2-oxazoline	14.9 ± 2.8	84.4 ± 0.5
2-Nonyl-2-oxazoline	7.58 ± 1.15	76.3 ± 0.5

Reprinted from Wiesbrock, F., Hoogenboom, R., Leenen, M.A.M., Meier, M.A.R., Schubert, U.S. 2005a. Macromolecules 38:5025, with permission.

ization were maintained. Moreover, the polymerization can be carried out in concentrated solutions or even bulk conditions to afford well-defined monomers (PDI < 1.20). Under such conditions, a maximum number of 300 monomers can be incorporated into the polymer chains.

Later, the same technique of the ring-opening polymerization under microwave irradiation was applied to the synthesis of a library of diblock copoly(2-oxazoline)s, in which a total number of 100 (50 + 50) monomer units were incorporated into the polymer chains (Wiesbrock et al., 2005b). As the result, 16 polymers were obtained with narrow PDI (less than 1.30) (Table 3.12). The reactions were initiated by methyltosylate and carried out in an acetonitrile solution at 140 °C. After polymerization of the first monomer, the reaction vessels were retransferred to an inert atmosphere of argon, the second monomer was added, and the reaction mixture was again irradiated in a microwave reactor.

Very recently, the same method was successfully used for the preparation of triblock copolymers under microwave irradiation (Hoogenboom et al., 2006). A library of 30 triblock copolymers was prepared from 2-methyl-, 2-ethyl-, 2-nonyl-, and 2-phenyl-2-oxazoline in a single-mode microwave reactor. The polymers exhibited narrow molecular weight distributions (PDI < 1.33) and showed only minor deviations from the targeted monomer ratio of 33:33:33 (Table 3.13).

The glass transition temperature of the triblock copolymers spanned the range from 50 to 100 °C depending on the incorporated monomers. The micellization study in water of selected triblock copolymers consisting of two water-soluble segments and one hydrophobic segment revealed a clear effect of the block order on the micellar size. The micelles from polymers with a hydrophobic middle block were smaller compared with micelles from polymers with end-standing hydrophobic segments (Table 3.14) (Hoogenboom et al., 2006)

Table 3.12. (Theoretical) number average molecular weight M_n^{th} (kDa) and polydispersity index (PDI) for the four chain-extended and the 16 diblock copoly(2-oxazoline)s*.

First monomer	Second monomer			
	2-Methyl-2-oxazoline	2-Ethyl-2-oxazoline	2-Nonyl-2-oxazoline	2-Phenyl-2-oxazoline
2-Methyl-2-oxazoline	M_n^{th} = 8.5 PDI = —/1.16	M_n^{th} = 9.2 PDI = —/1.17	M_n^{th} = 14.2 PDI = —/—	M_n^{th} = 11.6 PDI = —/1.25
2-Ethyl-2-oxazoline	M_n^{th} = 9.2 PDI = —/1.18	M_n^{th} = 9.9 PDI = 1.12/1.16	M_n^{th} = 14.8 PDI = 1.15/—	M_n^{th} = 12.3 PDI = 1.27/1.19
2-Nonyl-2-oxazoline	M_n^{th} = 14.2 PDI = —/—	M_n^{th} = 14.8 PDI = 1.64/—	M_n^{th} = 19.7 PDI = 1.14/—	M_n^{th} = 17.2 PDI = 1.24/—
2-Phenyl-2-oxazoline	M_n^{th} = 11.6 PDI = —/1.18	M_n^{th} = 12.3 PDI = 1.35/1.19	M_n^{th} = 17.2 PDI = 1.28/—	M_n^{th} = 14.7 PDI = 1.27/1.16

*In each cell, the first (second) entry for the polydispersity indices results from measurements in chloroform (*N,N*-dimethylformamide).

Reprinted from Wiesbrock, F., Hoogenboom, R., Van Nispen, S.F.G.M., van Nispen, M., van der Loop, M., Abeln, C.H., van den Berg, A.M.J., Schubert, U.S. 2005b. Macromolecules 38:7957, with permission.

Step-Growth Polymerization

A number of polymers obtained in step-growth polymerization reactions can be accounted to so-called functionally terminated thermoplastics, which combine toughness of thermoplastics and resistance of thermosets. The challenge in the microwave synthesis and processing of these polymers is that temperature during the reactions and processing is often very close to their thermal degradation temperature, making temperature control crucial.

Polyethers and Polyesters

Synthesis of a polyether by phase-transfer catalysis (PTC) under microwave irradiation was investigated for the reaction of 3,3-bis(chloromethyl)oxe-tane (BCMO) and various bisphenols (i.e., bisphenol A, 4,4′-dihydroxya-zobenzene, 4,4′-dihydroxybiphenyl, 4,4′-dihydroxybenzophenone, and 4,4′-thiodiphenyl) to the synthesis (Fig. 3.29) (Hurduc et al., 1997).

In a typical reaction, a mixture of bisphenol (1.3 mmol), water (5 mL), NaOH (3 g), nitrobenzene (5 mL), BCMO (1.3 mmol), and TBAB (0.2 mmol) was placed in a 50-mL flask and irradiated for 1.5 hours in a microwave waveguide (60 W), while temperature (95 to 100 °C) was monitored by a thermovision infrared camera. Under conventional conditions,

Table 3.13. Number of incorporated monomer units into the 30 triblock copoly(2-oxazoline)s resulting from combined ^{1}H NMR analyses (top) of the model [A and AB (block-co)polymers] and final polymers as well as the measured number average molecular weights ($M_{n,GPC}$/PDI; bottom)*.

First-second block	Third block			
	MeO$_X$	**EtO$_X$**	**PhO$_X$**	**NoO$_X$**
MeO$_X$-EtO$_X$	33:28:33 10.2 kDa/1.21		33:33:32 11.7 kDa/1.24	33:31:33 5.5 kDa/1.44
MeO$_X$-PhO$_X$	33:31:33 14.1 kDa/1.22	33:33:36 13.9 kDa/1.15		33:30:32 10.2 kDa/1.21
MeO$_X$-NoO$_X$	33:28:33 9.9 kDa/1.20	33:30:37 10.0 kDa/1.21	33:29:29 10.6 kDa/1.27	
EtO$_X$-MeO$_X$		33:33:33 10.9 kDa/1.32	33:29:27 12.4 kDa/1.23	33:34:31 9.5 kDa/1.28
EtO$_X$-PhO$_X$	33:31:30 16.2 kDa/1.20	33:30:33 15.3 kDa/1.24		33:30:36 11.4 kDa/1.22
EtO$_X$-NoO$_X$	33:33:37 10.1 kDa/1.27	33:33:33 9.9 kDa/1.22	33:33:31 11.3 kDa/1.25	
PhO$_X$-MeO$_X$		33:35:35 15.3 kDa/1.21	33:27:33 15.2 kDa/1.19	33:31:31 9.1 kDa/1.23
PhO$_X$-EtO$_X$	33:35:34 17.2 kDa/1.32		33:42:33 19.1 kDa/1.28	33:38:38 14.1 kDa/1.21
PhO$_X$-NoO$_X$	33:38:34 9.7 kDa/1.21	33:45:37 8.8 kDa/1.21	33:36:33 11.6 kDa/1.22	
NoO$_X$-PhO$_X$	33:23:27 7.2 kDa/1.40	33:26:24 7.8 kDa/1.33		33:32:33 10.3 kDa/1.38†

*^{1}H NMR spectra were recorded in CDCl$_3$ or CD$_2$Cl$_2$ (PhO$_X$-containing polymers), and GPC analyses were performed using DMF (with 5 mM NH$_4$PF$_6$) as eluent. $M_{n,GPC}$ was calculated utilizing PMMA standards.

†GPC measurement with CHCl$_3$:Net$_3$:2-PrOH (94:4:2) as eluent (PS calibration).

Reprinted from Hoogenboom, R., Wiesbrock, F., Huang, H., Leenen, M.A.M., Thijs, H.M.L., van Nispen, S.F.G.M., van der Loop, M., Fustin, C.A., Jonas, A.M. Gohy, J.F., Schubert, U.S. 2006. Macromolecules 39:4719, with permission.

the reaction was carried out for 5 hours at 90 °C. It was found that for semicrystalline polymers, the yields were higher under microwave conditions, whereas in the case of amorphous polymer, the yields were approximately equal plus being of shorter reaction time.

The synthesis of linear polyethers from either isosorbide (Chatti et al., 2002) or isoidide (Chatti et al., 2003) and disubstituted alkyl bromides or methanesulfates by using microwave irradiation under solid-liquid PTC conditions was also described (Fig. 3.30) (Chatti et al., 2002, 2003).

Table 3.14. Characteristic features of the micelles formed by the six copolymers investigated in this study.

Triblock copolymer	Property		
	C (g × L⁻¹)	R$_h$(nm)/PDI	D$_z$ (nm)
p(EtO$_X$-b-MeO$_X$-b-PhO$_X$)	0.19	19/0.29	18 ± 2.3
p(MeO$_X$-b-EtO$_X$-b-PhO$_X$)	0.21	25/0.49	16 ± 2.2
p(EtO$_X$-b-PhO$_X$-b-MeO$_X$)	0.18	13/0.52	7 ± 1.7
p(EtO$_X$-b-MeO$_X$-b-NonO$_X$)	0.20		10 ± 1.8
p(MeO$_X$-b-EtO$_X$-b-NonO$_X$)	0.19	36/0.34	13 ± 2.0
p(EtO$_X$-b-NonO$_X$-b-MeO$_X$)	0.18	20/0.44	7 ± 1.3

C is the concentration of the micelles in pure water after the dialysis step. R_h is the hydrodynamic radius (obtained from the CONTIN histogram for the individual micelles. D_z are obtained from AFM images. Note that the CONTIN analysis of the DLS data for p(MeO$_X$-b-EtO$_X$-b-NonO$_X$) could not resolve a population associated with micelles.

Reprinted from Hoogenboom, R., Wiesbrock, F., Huang, H., Leenen, M.A.M., Thijs, H.M.L., van Nispen, S.F.G.M., van der Loop, M., Fustin, C.A., Jonas, A.M. Gohy, J.F., Schubert, U.S. 2006. Macromolecules 39:4719, with permission.

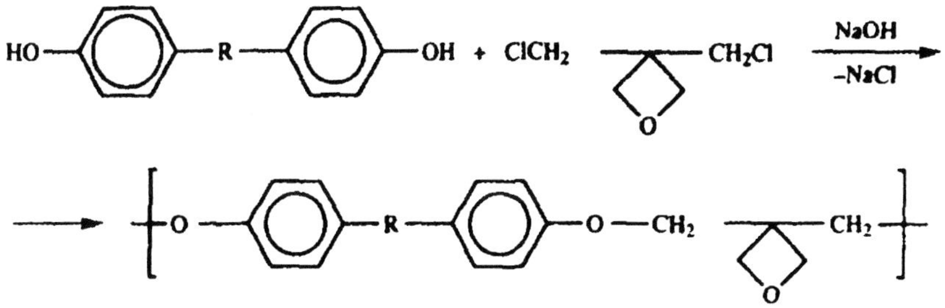

Fig. 3.29 Synthesis of polyethers by phase-transfer catalysis (PTC) reactions.

Isosorbide and isoidide are byproducts of biomass obtained from the sugar industry through double dehydration of starch (Fleche et al., 1986). The reactions were carried out in a single-mode microwave reactor Synth-wave 402 (Prolabo) with temperature infrared detector, which was previously calibrated with optical fiber detector introduced into a reaction mixture. The reaction mixtures consisting of isosorbide or isoidide (5 mmol), alkyl dibromide/methanedisulfate (5 mmol), tetrebutylammonium bromide (TBAB, 1.25 mmol) and powdered KOH (12.5 mmol) were irradiated for 30 minutes to afford the polyethers with 70% to 90% yield (Fig. 3.31). It was

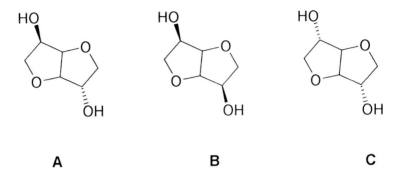

Fig. 3.30 (**A**) Isosorbide (1,4-dianhydro-D-sorbitol). (**B**) Isomannide (1,4-dianhydro-D-mannitol). (**C**) Isoidide (1,4-dianhydro-D-iditol).

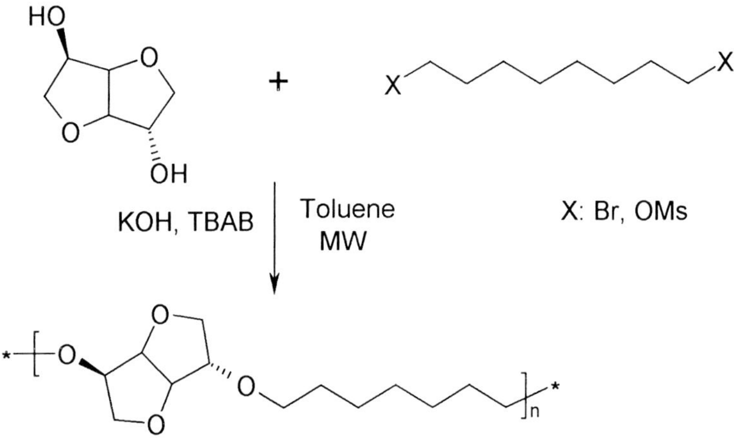

Fig. 3.31 Synthesis of linear polyethers from either isosorbide or isoidide and disubstituted alkyl bromides or methanesulfates.

found that the use of a small amount of solvent was necessary to ensure good temperature control and decrease in the viscosity of the reaction medium.

In the case of isosorbide, the microwave-assisted synthesis proceeded more rapidly, compared with conventional heating, and was reduced to 30 minutes with the yield of approximately 69% to 78%. Under conventional conditions, the polyethers were afforded with 28% to 30% yield within 30 minutes. Similar yields of the polyethers were obtained when the reaction

Table 3.15. Influence of reaction time on the yields of high-molecular-weight fraction (FP MeOH) and low-molecular-weight fraction (FP Hex) of polyethers.

Time	Mode of activation	FP MeOH (%)	FP Hex (%)	Total yield (%)
30 min	M	67	18	85
60 min	MW	71	19	90
30 min	Δ	12	81	93
1 day	Δ	64	25	89
1 week	Δ	83	5	88
1 month	Δ	91	0	91

MW, microwave irradiation; Δ, conventional thermal heating.
Reprinted from Chatti, S., Bortolussi, M., Loupy, A., Blais, J.C., Bogdal, D., Roger, P. 2003. J. Appl. Polym. Sci. 90:1255, with permission

time was extended to 24 hours. These yields remained practically unchanged even though the synthesis was carried out for another 7 days (Table 3.15).

The analysis of properties of the synthesized polyethers revealed that the structure of the products strictly depended on the activation mode (i.e., microwave or conventional activation). Under microwave conditions, the polyethers were characterized by higher molecular weight and better homogeneity. For example, within 30 minutes of the reaction time under conventional heating, the polyethers with higher molecular weight were not observed at all. Moreover, it was found that the mechanism of chain termination is different under microwave and conventional conditions. The polyethers prepared with conventional heating have shorter chains with terminal hydroxyl ends, whereas under microwave irradiation the polymer chains were longer with terminal ethylenic ends. In fact, under microwave irradiation, terminal ethylenic ends were formed rapidly and set up a hindrance to further polymer growth. In opposite, under conventional conditions the terminations were essentially constituted by hydroxyl functions; however, the further polymerization was terminated as well (Chatti et al., 2002).

Later, the same protocol was applied to the polycondensation of aliphatic diols of isosorbide with 1,8-dimesyloctane and other dibromo- and disulfonated alkylating agents (Fig. 3.32) (Chatti et al., 2004) In all the cases, it was found that microwave-assisted polycondensations proceeded more efficiently compared with conventional heating (the reaction time was reduced from 24 hours to 30 minutes: ratio 1:50). The polycondensation under microwave yields 63% of polyethers with relatively high average-weight molecular weights (weight average molecular weight up to 7,000

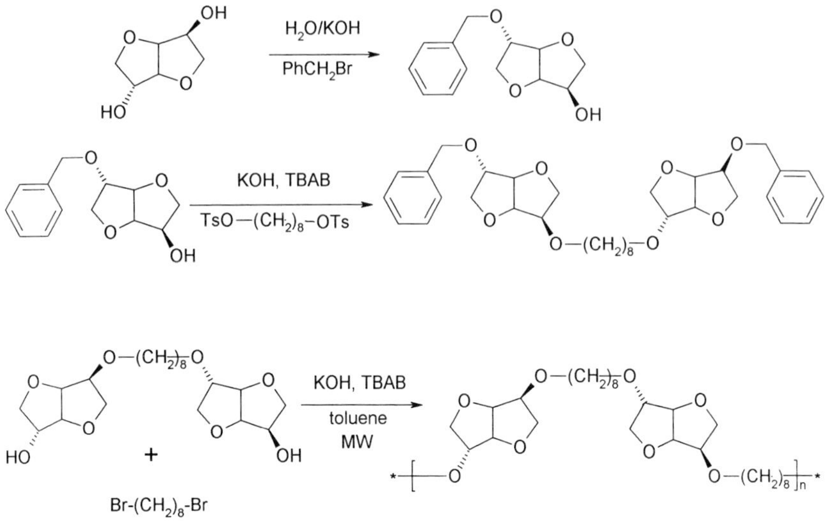

Fig. 3.32 Polycondensation of aliphatic diols of isosorbide with 1,8-dimesyloctane and other dibromo- and disulfonated alkylating.

g/mol). The polyethers were characterized by NMR (^{1}H, ^{13}C) and Fourier transform–infrared spectroscopy and SEC measurement and MALDI-TOF mass spectrometry.

It was demonstrated that the application of previously synthesized ethers of isosorbide was beneficial and allowed preparation of polyethers in better yields than the polyethers obtained in direct reactions of isosorbide and dibromo- or dimesylalkenes. Moreover, the molecular weights of the polyethers were higher than of those prepared in the earlier work, while molecular weight distributions of new polyethers were similar or lower. Such a microwave-assisted procedure can contribute to synthesis of alternating polyethers and further modification of their properties. Noticeably different results that were obtained when polycondensation reaction was performed under microwave irradiation instead of classic heating were ascribed to the following (Chatti et al., 2004):

- Reaction times were substantially decreased as already advocated for such type of alkylation. This effect can be attributed to an increase in system polarity from the ground state (GS) of the reaction, which consists in tight ions pairs (involving hard alkoxide anions) toward the transition where ions pairs are much looser due to negative charge delocalization. It results in an enhancement in polarity during the reaction progress and, consequently, to an increase of microwave–

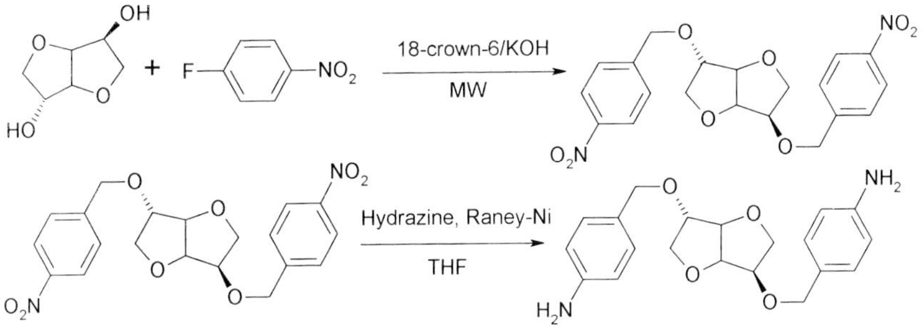

Fig. 3.33 Reaction of isosorbide with *p*-fluoronitrobenzene and further reduction of the dinitro derivative into the diamino derivative.

materials interactions magnitude responsible for the observed acceleration.

- Higher molecular weight polymers were obtained with a better molecular weight distribution. This effect could originate from some dipole orientation influence in the presence of an electromagnetic field that could induce some mutual special orientation of reagents. Such a conclusion was already drawn from some studies in the literature.

- The structure of polymers are quite different with essentially large discrepancies in the chain terminations as shown by the MALDI-TOF mass spectra. In fact, under microwave irradiation, ethylenic end groups are formed rather rapidly and set up a hindrance to further growth of polymers. Reciprocally, under conventional heating, terminations are essentially constituted by hydroxyl functions that nevertheless do not allow further polycondensation. Evidently, ethylenic formation appeared as favored under microwave. This observation is consistent with an enhanced microwave stabilization by dipole–dipole interaction of the more polar transition state (TS). In elimination TS, due to a greater delocalization of negative charge compared with substitution TS, polarity is enhanced in connection with a looser ion pair.

Recently, a novel group of polyamides were synthesized by the microwave-assisted polycondensation of an optically active isosorbide-derived diamine with different diacyl chlorides in the presence of a small amount of *N*-methyl-2-pyrrolidone (NMP) (Caouthar et al., 2005). For this purpose, a new diamine derivative of isosorbide was prepared by the reaction of isosorbide with *p*-fluoronitrobenzene and further reduction of the dinitro derivative into the diamino derivative under microwave irradiation (Fig. 3.33).

The polymers were synthesized by the reaction of acid dichlorides and the diamino derivative of isosorbide under microwave conditions. The re-

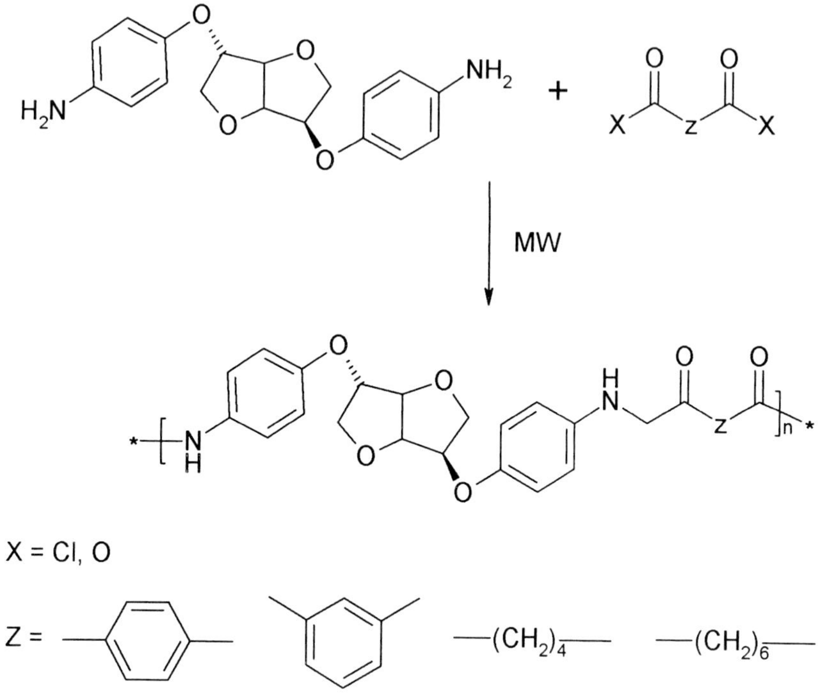

Fig. 3.34 Polymerization of acid dichlorides and the diamino derivative of isosrbide.

actions were carried out in a cylindrical vessel fitted to a single-mode microwave reactor, in which 1 mmol of the diamine derivative was mixed with 1 mmol of diacyl chloride dissolved in 1 mL of NMP. The reaction mixture was mechanically stirred during the whole irradiation period (6 minutes at 200 °C). After cooling to room temperature, the polyamide was obtained by precipitation in methanol (Fig. 3.34).

The use of such conditions was necessary to induce effective homogeneous heating of the monomers under microwave irradiation and subsequent polycondensation leading to the formation of the polyamide with inherent viscosities between 0.22 and 0.72 dL/g that corresponded to rather higher molecular weights up to 140,000 g/mol. Applying interfacial polymerization under conventional heating conditions, lower-molecular-weight polyamides were obtained with inherent viscosities in the range of 0.04 to 0.36 dL/g that was related to the maximum molecular weight of approximately 29,000 g/mol (Caouthar et al., 2005).

Unsaturated polyesters were obtained in the polyaddition reactions of alkylene oxides (i.e., epichlorohydrin) and acid anhydrides (i.e., maleic and

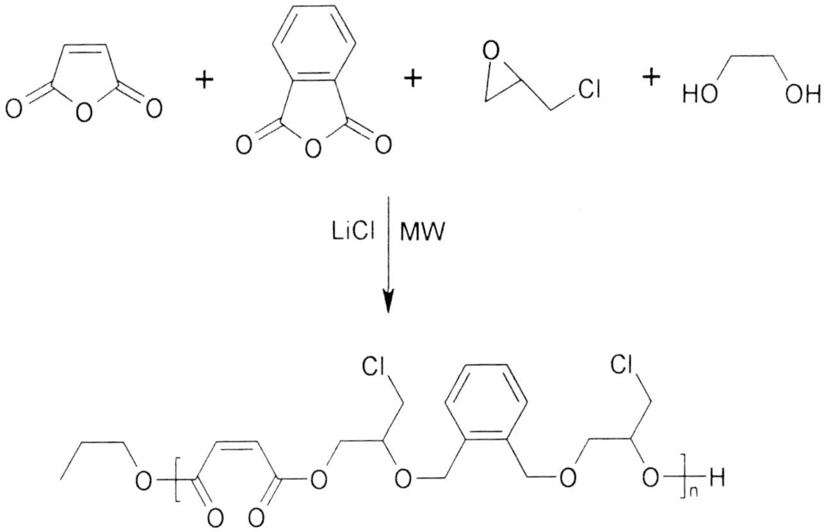

Fig. 3.35 Synthesis of polyesters in the polyaddition reactions of alkylene oxides (i.e., epichlorohydrin) and acid anhydrides.

phthalic anhydrides) in the presence of lithium chloride as a catalyst under microwave irradiation (Fig. 3.35) (Pielichowski et al., 2003a, 2004).

In a standard procedure, a mixture of 0.10 mole of phthalic anhydride and 0.10 mole of maleic anhydride with 0.010 mole of ethylene glycol (EG), 0.20 mole of epichlorohydrin (EPI), and lithium chloride (0.1 wt%) was placed in a three-necked round-bottomed flask and irradiated in a multimode microwave reactor equipped with a temperature infrared sensor, magnetic stirrer, and upright condenser. The reaction temperature was maintained at the range of 120 to 140 °C under an inert atmosphere. The polymerization was continued until acid number value of polyesters dropped below 50 mg KOH/g. Compared with polycondensation reactions of acid anhydrides with diols, these reactions proceed without a release of any byproducts. At the same time, the polyaddition reactions were performed under conventional thermal conditions, applying a similar set of reaction conditions. In comparison with these experiments, a 2-fold reduction of reaction times was observed under microwave conditions while other parameters such as number average molecular weight and PDI were comparable (Table 3.16).

The comparison polyaddition reactions were carried out in multimode and single mode microwave reactors with microwave power irradiation of 600 and 300 W, respectively (Table 3.17) (Pielichowski et al., 2004). It can be seen that the choice of the microwave reactor type influenced molecular weights of unsaturated polyesters while acid number value of polyesters

Table 3.16.   Characteristics of unsaturated polyester resins obtained under microwave irradiation and under conventional heating.

Temp., °C	Reaction time, min	EPI/ anhydrides/ EG, molar ratio	M_n	M_w	M_w/M_n	Acid number, mg KOH/g	Gardner scale color
		Microwave irradiation					
120	120	2.0:2.0:0.1	1541	3274	2.13	43	3
120	120	2.2:2.0:0.1	1954	3685	1.89	20	3
130	100	2.0:2.0:0.1	2197	5524	2.51	28	3
130	90	2.2:2.0:0.1	1566	5795	3.70	14	3
140	60	2.0:2.0:0.1	2019	8513	4.22	27	4
140	45	2.2:2.0:0.1	2179	5386	2.47	47	4
		Thermal heating					
120	300	2.0:2.0:0.1	2104	4021	1.91	17	2
120	300	2.2:2.0:0.1	2156	3969	1.84	19	2
130	180	2.0:2.0:0.1	3125	10,900	3.49	12	4
130	180	2.2:2.0:0.1	3645	14,000	3.84	2	4
140	120	2.0:2.0:0.1	1931	7413	3.84	34	4
140	90	2.2:2.0:0.1	2267	6445	2.84	8	3

Reprinted from Pielichowski, J., Penczek, P., Bogdal, D., Wolff, E., Gorczyk, J. 2004. Polimery 49:763, with permission.

Table 3.17.   Characteristics of unsaturated polyester resins obtained under monomode and multimode microwave irradiation.

mw system	Temp., °C	Reaction time, min	EPI/ anhydrides/ EG, molar ratio	M_n	M_w	M_w/M_n	Acid number, mg KOH/g
Multimode	130	100	2.0:2.0:0.1	2197	5524	2.51	28
Monomode	130	120	2.0:2.0:0.1	1200	3000	2.50	30
Multimode	120	120	2.2:2.0:0.1	1954	3685	1.89	20
Monomode	120	120	2.2:2.0:0.1	1580	3360	2.32	9

Reprinted from Pielichowski, J., Penczek, P., Bogdal, D., Wolff, E., Gorczyk, J. 2004. Polimery 49:763, with permission.

in case of both reactors reached the desired level (i.e., less than 50 mg KOH/g). It was found that higher molecular weights can be obtained in the multimode reactor. Moreover, the multimode system can offer more flexibility and control during the design of microwave experiments.

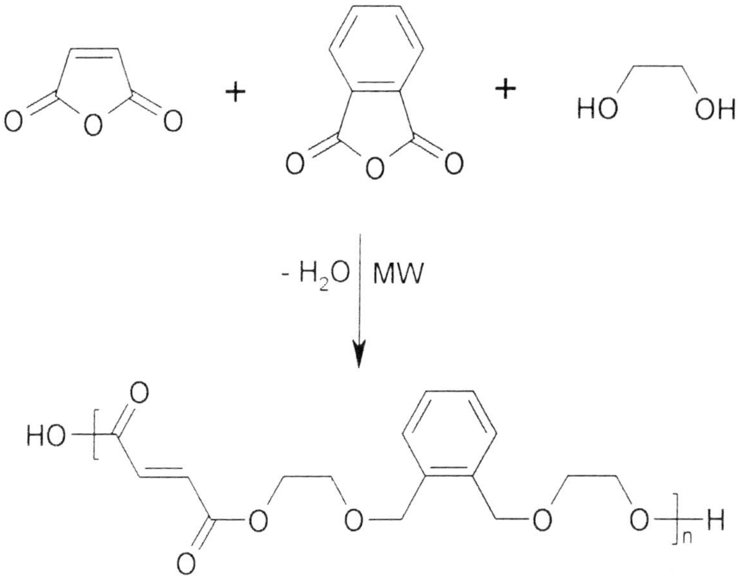

Fig. 3.36 Polycondesantion of acid anhydrides (i.e., maleic and phthalic anhydrides) with diols.

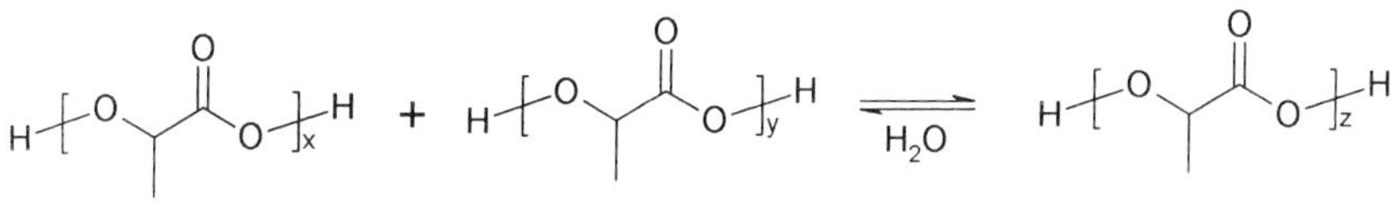

Fig. 3.37 Polycondensation of DL-lactic acid (LA).

In turn, polycondensation of acid anhydrides (i.e., maleic and phthalic anhydrides) with diols (i.e., ethylene glycol) under microwave irradiation was also described for the synthesis of unsaturated polyesters (Pielichowski et al., 2003a). In addition to the previous protocol, the reaction temperature was increased to 200 °C, and a Dean-Stark trap was applied to remove water from the reaction mixture (Fig. 3.36). It was found that the reaction times for microwave and conventional protocols were comparable and depended on the rate of removing water from the reaction system.

The bulk polycondensation of DL-lactic acid (LA) to poly(lactic acid) under microwave irradiation was studied as well (Fig. 3.37) (Keki et al., 2001). The reactions were carried out in a domestic microwave oven, in which 5 g of LA were placed in a 20-mL beaker and irradiated at 650 W. The reaction time was varied between 10, 20, and 30 minutes to afford poly(lactic acid) with 96%, 84%, and 63% yield, respectively.

The yield of polycondensation decreased with increasing irradiation time, probably due to the loss of oligomers of lower polymerization degree during the polycondensation. Under conventional thermal conditions, the polymer was obtained in 94% yield after 24 hours at 100 °C. According to MALDI MS analysis, only linear oligomers were formed. After 20 minutes of microwave irradiation, oligomers with nearly the same molecular weight were obtained like those produced upon conventional heating (i.e., 100 °C, 24 hours). Thus, the reaction time for preparing poly(lactic acid) could be considerably reduced. On increase in the reaction time under microwave conditions, the fraction of cyclic oligomers tended to appear beside linear ones (Keki et al., 2001).

In turn, the synthesis of polyesters by the step-growth polymerization reaction under microwave irradiation was presented for polycondensation of 1,4-butanediole and succinic acid in the presence of 1,3-dichloro-1,1,3,3-tetrabutyldistannoxane as a catalyst (Fig. 3.38) (Velmathi et al., 2005).

Typically, 1,4-butanediol (5 mmol), succinic acid (5 mmol) and the catalyst (0.1 mmol) were placed in a flask and irradiated in single-mode microwave reactor. The mixture was stirred at 200 °C under nitrogen atmosphere at maximum power of 200 W. The polymer formed [i.e., poly(butylene succinate)(PBS)] was dissolved in chloroform and precipitated into methanol. In the case of solution polymerization, decalin was added (5 mmol of reactants for 10 mL of decaline) to the reaction mixture, which allowed for trapping water using a Dean-Stark apparatus. For comparison, the polymerization under conventional conditions was carried out for 5 hours in an oil bath preheated to 200 °C. PBS was dissolved in chloroform and precipitated into methanol. It was found that under microwave irradiation, PBS was achieved within 20 minutes with weight average molecular weight (M_w) of 1.03×10^4 g/mol, while under conventional conditions, a similar value of weight average molecular weight (M_w 1.02×10^4 g/mol) was obtained after 5 hours. The kinetic results of the synthesis of PBS by conventional heating and microwave irradiation are presented in the Figure 3.39 (Velmathi et al., 2005).

Polyamides

Synthesis of aliphatic polyamides under microwave irradiation was reported in a series of papers (Imai, 1996a, 1996b, 1999; Imai et al., 1996a, 1996b). Polyamides were prepared of both ω-amino acids and diamines together with dicarboxylic acids (i.e., the nylon salt type of monomers) in the presence of a small amount of a polar organic medium (Fig. 3.40).

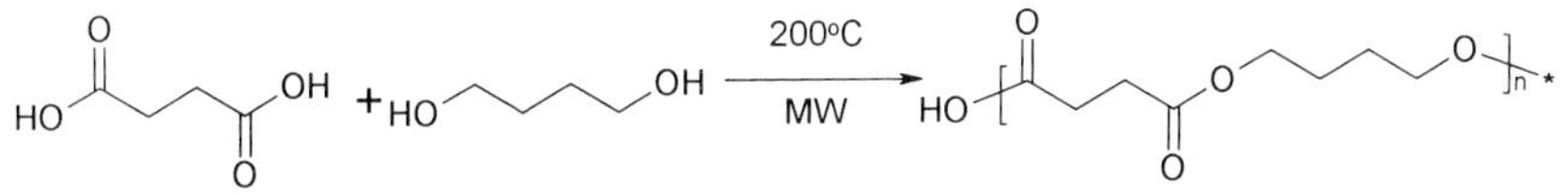

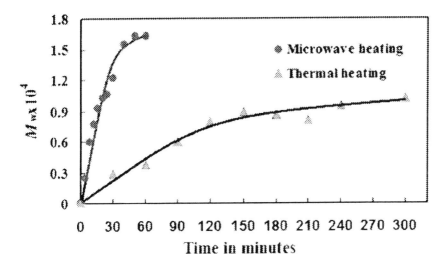

Fig. 3.38 Polycondensation of 1,4-butanediole and succinic acid in the presence of 1,3-dichloro-1,1,3,3-tetrabutyldistannoxane as a catalyst.

Fig. 3.39 Kinetic of PBS synthesis under microwave and conventional conditions. Reprinted from Velmati, S., Nagahata, R., Sudiyama, J., Takeuchi, K. 2005. Macromol. Rapid. Commun. 26:1163.

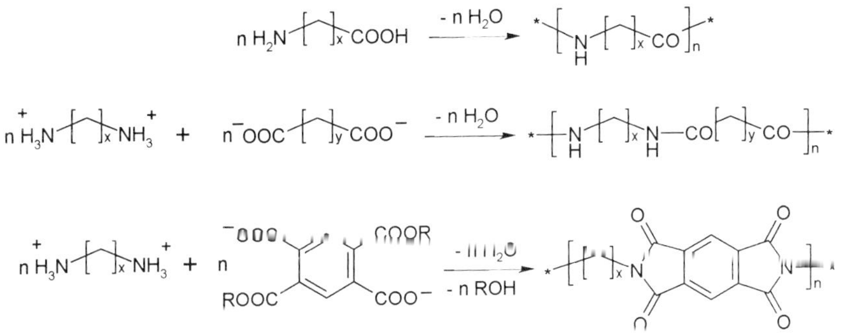

Fig. 3.40 Synthesis of aliphatic polyamides from ω-amino acids as well as diamines and dicarboxylic acids.

Table 3.18.   Solvent effect on microwave-assisted polycondensation of 12-aminododecanoic acid.[a]

Solvent type	Solvent ε	Solvent bp, °C	Ft,[b] °C	Reaction Solvent time, min	Ft,[c] °C	Polymer η_{inh},[d] dL × g^{-1}
Water	78	100	97	5	220	0.35
Dimethyl sulfoxide	47	189	172	4	259	0.24
Dimethylacetamide	38	166	163	5	281	0.23
N-Methylpyrrolidone	32	202	179	4	267	0.39
Nitrobenzene	35	211	198	5	264	0.42
Ethanediol	38	197	193	5	317	0.59
1,4-Butanediol	31	229	189	5	242	0.24
Diphenyl ether	4	258	66	5	109	—

[a] The polymerization was carried out with 2 g of monomer and 2 mL of the solvent under microwave irradiation.
[b] Final temperature of the solvent alone after 2 minutes of microwave irradiation.
[c] Final temperature of reaction mixture.
[d] Measured at a concentration of 0.5 d dl^{-1} in *m*-cresol at 30 °C.
Reprinted from Imai, Y., Nemoto, H., Watanabe, S., Kakimoto, M. 1996b. Polymer. J. 28:256, with permission.

The reactions were carried out in a modified domestic microwave oven with a small hole on the top of the oven so that nitrogen was introduced to a 30-mL wide-mouth vial adapted as a reaction vessel. In a typical experiment, a monomer or its salt (2 g) in a polar high-boiling solvent (1 to 2 mL) that acted as a microwave absorber was irradiated under nitrogen atmosphere. The microwave-assisted polycondensation proceeded rapidly and was completed within 5 minutes for the polyamides with inherent viscosity of around 0.24 to 0.63 dL/g (Imai et al., 1996a). For example, the solvent effect on the inherent viscosity of the polyamide formed by the polycondensation of 12-aminododecanooic acid for microwave irradiation is summarized in Table 3.18.

Large temperature differences were observed between the final temperature of the solvent alone and the final temperature of the polymerization mixtures. However, the temperature measurements were performed by means of a thermocouple that was immersed in the reaction mixtures just after removal of the reaction vessels from the microwave cavity. This phenomenon was explained by the participation of the polar amino acid monomer dissolved in the solvent to generate additional heat during the polymerization under microwave irradiation. When water, which possesses the highest dielectric constants, was used as a solvent, the polyamide with an inherent viscosity of 0.35 dL/g was obtained. In general, the use of the solvents possessing both a high dielectric constant and boiling point, attaining high final polymerization temperature, led to the formation of a polyamide characterized by viscosities higher than 0.5 dL/g. Suitable solvents were found to be sulfolane and amide-type solvents such as *N*-cyclohexyl-2-pyrrolidone and 1,3-dimethylimidazolidone. Hydroxyl-containing solvents possessing a high dielectric constant like glycol, and those with low polarity, such as benzyl alcohol, *m*-cresol, and *o*-chlorophenol, were also effective in producing a polyamide with the viscosity values of approximately 0.5 dL/g or higher. Thus, the solvents played an important role as primary microwave absorbers for the microwave-assisted polycondensations (Imai et al., 1996a).

The same experimental procedure was applied and similar observations were made for the polymerization of diamines with dicarboxylic acids (i.e., the nylon-type salt monomers). Moreover, for the salt monomers, the rate of polycondensation under various conditions was in the following order: microwave-induced polycondensation > solid-state thermal polymerization > high-pressure thermal polycondensation (Imai et al., 1999b). For some series of diamines and dicarboxylic acids, the polycondensation reactions were run under periodic and continuous microwave irradiation (Watanabe et al., 1993). Eventually, it was found that the polycondensation under periodic microwave irradiation yielded polyamides with a higher inherent viscosity and allowed for easier temperature control in comparison with continuous irradiation.

In a similar way, the synthesis of aromatic polyamides from aromatic diamines [*m*-phenylenediamine, *p*-phenylenediamine, bis(4-aminophenyl) methane, and bis(4-aminophenyl)ether] and dicarboxylic acids such as isophthalic and terephthalic was performed in a household microwave oven (Park et al., 1993). The polycondensation was carried out in a NMP solution in the presence of triphenylphosphite (TPP), pyridine, and lithium chloride as condensing agents to produce a series of polyamides with moderate inherent viscosities of 0.21 to 0.92 dL/g within 30 to 50 seconds. However, no marked differences in molecular weight distribution and inherent viscosities were found between the polyamides produced by conventional (60 seconds, 220 °C) and microwave methods (Park et al., 1993).

Recently, the synthesis of polyamides from linear nonaromatic dicarboxylic acids (i.e., adipic, suberic, sebacic, and fumaric acid) and aromatic diamines such as *p*-phenylenediamine or 2,5-bis(4-aminophenyl)-3,4-diphenylthiophene (Fig. 3.41) under microwave conditions was presented (Pourjavadi et al., 1999).

The reactions were carried out in a 50-mL polyethylene (HDPE) screw-capped cylinder vessel, in which aromatic acid (1.25 mmol) with aliphatic diamine (1.25 mmol) in a NMP (3 mL) solution were irradiated in a domestic microwave oven (30 to 40 seconds) in the presence of TPP (3.123

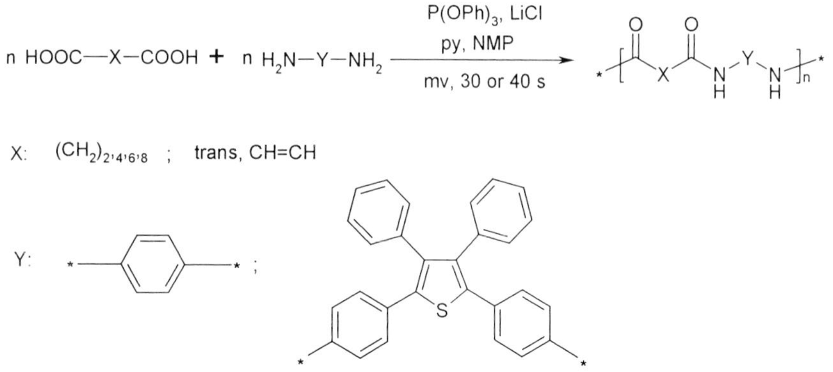

Fig. 3.41 Synthesis of polyamides from linear nonaromatic dicarboxylic acids (i.e., adipic, suberic, sebacic, and fumaric acid) and aromatic diamines.

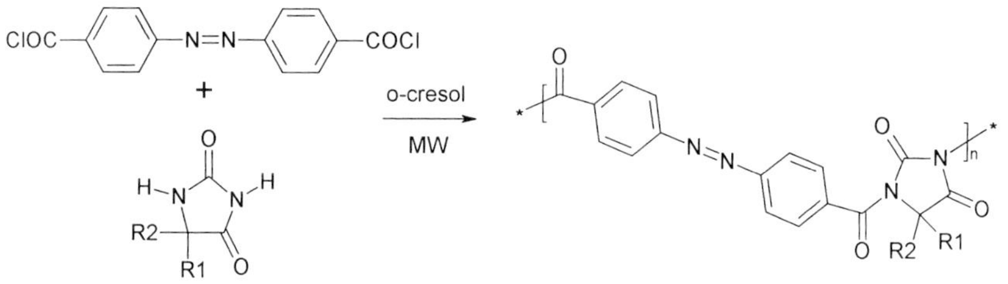

Fig. 3.42 Synthesis of polyamides containing azobenzene units and hydantoin derivatives.

mmol), pyridine (0.75 mL), and lithium chloride (3.123 mL). The polyamides with inherent viscosity in the range of 0.1 to 0.8 dL/g were obtained in medium to high yield (60% to 100%) (Pourjavadi et al., 1999). Temperature was not detected during these microwave experiments. The polyamides were characterized by thermal methods (TGA, DSC). However, no comparison to the polymers prepared by a conventional method was made.

More recently, the synthesis of polyamides containing azobenzene units and hydantoin derivatives in the main chains under microwave irradiation was described (Faghihi et al., 2003). The polycondensation of 4,4'-azodibenzoyl chloride with eight different derivatives of 5,5-disubstituted hydantoin in the presence of a small amount of o-cresol is shown in Fig. 3.42.

The polycondensation reactions were carried out within 8 minutes in a domestic microwave oven in a porcelain dish in which 1.0 mmol of diacid chloride was mixed with an equimolar amount of diol in the presence of small amounts of o-cresol. The polymerization proceeded rapidly, com-

Table 3.19. Comparison of microwave and thermal activation in copolymerization reactions.

Sample	Δ-PAE, 1%[a] 160 °C- 0.5 h	Δ-PAE, 2% 160 °C- 0.5 h	Δ-PAE, 3% 160 °C- 0.5 h	MW-PAE, 1% 160 °C- 0.5 h	MW-PAE, 2% 160 °C- 0.5 h	MW-PAE, 3% 160 °C- 0.5 h
Starting materials ester/amide	1:2	1:2	1:2	1:2	1:2	1:2
Yield (%)	51.2	52.7	57.0	61.9	70.1	78.2
T_g (°C), (tan δ), DMTA (1 Hz)	−25.0	−18.5	−14.5	−14.0	−7.5	6.0
T_g (°C), from Fox equation	29	12	9	15	8	4
M_m (kg/mol), GPC	25.4	19.8	17.1	22.0	21.3	16.2
M_w/M_n	1.4	1.5	1.6	2.1	2.0	1.5

[a] Catalyst level.

Reprinted from Fang, X., Hutcheon, R., Scola, D.A. 2000b. J. Polym. Sci. Part A Polym. Chem. 38:1379, with permission.

pared with the bulk reactions under conventional conditions (8 minutes versus 1 hour), producing a series of the polyamides in high yield and inherent viscosity between 0.35 to 0.60 dL/g.

Microwave irradiation was also applied to synthesize poly(ε-caprolactam-co-ε-caprolactone) directly from the two cyclic monomers, ε-caprolactam and ε-caprolactone, by anionic catalyzed ring-opening polymerization (Fang et al., 2000b). The reactions were carried out using a variable-frequency (0.4 to 3 GHz) microwave oven, programmed for a set temperature and controlled by a pulsed power on-off system. During microwave experiments, ε-caprolactam and ε-caprolactone in the molar ratio of 2:1 were mixed together with solid $LiAl(OC(CH_3)_3)_3H$ (1 to 3 mol% of total reactants), and the mixture of 5 to 10 g was irradiated for 0.5 to 1 hour in a Teflon vessel under nitrogen blanket. The copolymerization temperature inside the sample (140 to 180 °C) was measured and controlled with a grounded thermocouple, which was calibrated against fiberoptic temperature probe. The same ratio of reactants and catalysts used in the microwave methods was used for the conventional thermal synthesis in an oil bath. Compared with the corresponding thermal products, microwave synthesized copolymers were obtained in higher yield, amide composition, and glass transition temperature and equivalent molecular weights (Table 3.19) (Fang et al., 2000b).

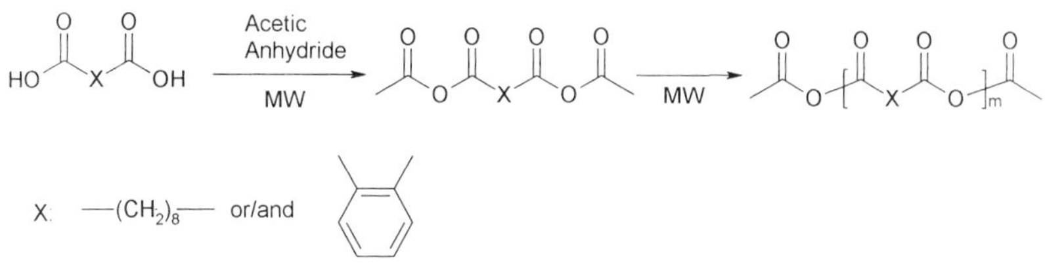

Fig. 3.43 Synthesis of polyanhydries prepared from aliphatic and aromatic diacids.

Table 3.20. Sebacic anhydride prepolymer polymerization with different catalysts under microwave irradiation. Reprinted from: Vogel B.M., Mallapragada, S.K., Narasimhan, B. (2004) Macromol. Rapid Commun. 25:330 with permission.

Catalyst	Time, min.	M_n, g/mol	DP
—	—	2017	11
SiO_2	4	1482	8
SiO_2	25	2112	11
Al_2O_3	25	5169	28
CaO	10	2929	16
Sand CaO	25	2363	13
Beads	25	11358	61
Beads	7	7542	41
Plate	5	1751	9

Miscellaneous Polymers

Polyanhydries that are counted among biodegradable polymers were also prepared from aliphatic and aromatic diacids under microwave irradiation (Fig. 3.43) (Vogel et al., 2004). The reactions were carried out in a domestic microwave oven.

The microwave experiments were broken down into two classes: first, reactions of prepolymers that directly produce polyanhydrides; second, in situ formation of prepolymers from diacid and polymerization under microwave irradiation (Table 3.20). For this purpose, 0.49 mmol of aliphatic or aromatic acid was mixed with 2.95 mmol of acetic anhydride and irradiated with full power in a sealed borosilicate vial for 2 minutes. Then the anhydride was evaporated, and the vial was irradiated for additional 5 to 25 minutes. In an attempt to increase the molecular weight of the polymers formed from prepolymers, catalysts such as calcium oxide, aluminum oxide, and silica were added, all of which have been used previously in conventional polymerizations to increase molecular weight. However, the

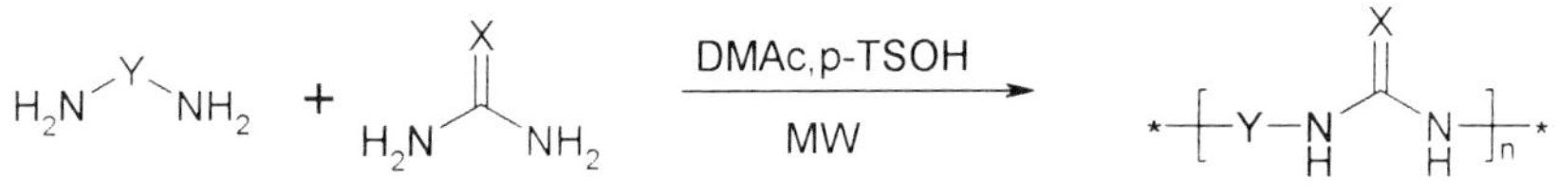

X = O, S   DMAc: N,N-dimethylacetamide   Y: —$(CH_2)_x$—   —Ar—

Fig. 3.44 Preparation of polyureas and polythioureas in the reaction of aromatic and aliphatic amines with urea and thiourea.

best results were obtained when the prepolymer was subjected to microwave irradiation on a glass plate or in a bath of glass beads. It was found that by applying this method it was possible to obtained polymers with number average weights (1,700 to 11,300 g/mol) comparable to those obtained under conventional conditions while decreasing reaction time from hours to 6 to 20 minutes. In addition, it was possible to prepare the copolymers of sebacic acid prepolymer and 1,6-bis-(*p*-carboxyphenoxy)hexane.

Polyureas and polythioureas were synthesized in the reaction of aromatic and aliphatic amines with urea and thiourea, respectively (Fig. 3.44) (Banihashemi et al., 2004). In a typical procedure, a mixture of 10 mmol of an amine, 10 mmol of urea, and a small amount of *p*-toluene sulfonic acid (1 mmol) in 5 mL of *N*,*N*-dimethylacetamide solution was irradiated for 7 minutes at 220 W and then for 8 minutes at 400 W in a tall beaker placed in a household microwave oven. As the result, a series of polyureas and polythioureas were obtained in a good yield and moderate inherent viscosity of 0.13 to 0.25 dL/g.

Synthesis of poly(aspartic acid)s (PAAs) under microwave irradiation was recently presented (Pielichowski et al., 2003b; Polaczek et al., 2003). PAAs can be prepared from L-aspartic acid thermally through a polycondensation procedure, optionally in the presence of an acid catalyst, to form poly(succinimide) (sometimes called anhydropolyaspartic acid), and then through hydrolysis with aqueous alkali metal hydroxide to form a solution of PAA metal salt. On the other hand, PAAs can be prepared from maleic anhydride instead of L-aspartic acid, which can reduce the cost of production (Fig. 3.45).

Both monomers (i.e., L-aspartic acid and maleic anhydride) were applied for the synthesis of PAAs under microwave conditions; however, prior to the reaction, maleic anhydride was converted into maleic acid ammonium salt in the reaction with ammonium hydroxide. Then, the reactions were carried out without a catalyst in propylene glycol solutions at a temperature range of 160 to 230 °C in a multimode microwave reactor. PAA was

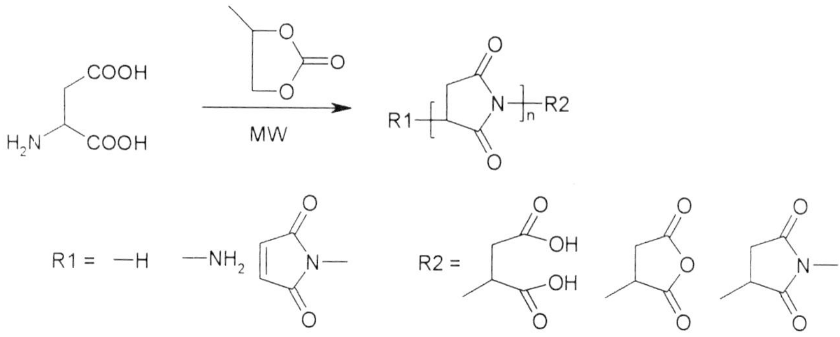

Fig. 3.45   Synthesis of poly(aspartic acid)s (PAAs).

Table 3.21.   Parameters of PAAs synthesis under thermal and microwave conditions from aspartic acid (KA) and maleic anhydride (SAM).

Monomer	Heating	Temperature (°C)	Reaction time (min)	Yield (%)
SAM	Δ	230	300	65
SAM	MW	230	20	93
KA	Δ	230	360	72
KA	MW	230	44	92

Reprinted from Polaczek, J., Pielichowski, J., Pielichowski, K., Tylek, E., Dziki, E. 2005. Polimery 50:812, with permission.

obtained in a good yield (50% to 85%) with number average molecular weights of 6,150 to 18,500 g/mol. For comparison, the synthesis of PAAs was carried out under conventional thermal heating conditions at the same temperature range as for microwave conditions. It was found that the application of microwave protocol allows significant reduction of reaction times from 5 hours to 20 minutes and from 6 hours to 40 minutes when PAAs are prepared from maleic anhydride and L-aspartic acid, respectively (Table 3.21) (Polaczek et al., 2005).

Polyarylene-type polymer was prepared through the polymerization of 2,7-dibromo-9,9-dihexylfluorene under microwave irradiation (Fig. 3.46) (Carter, 2002). In a typical experimental procedure, a catalyst stock solution was prepared consisting of 704 mg of bis(1,5 cyclooctadiene)nickel(0), 410 mg of 2,2'-bipyridine, and 281 mg of cyclooctadiene in 17.5 g of toluene/DMF (1.15:1). Then, a monomer solution containing 600 mg of 2,7-dibromo-9,9-dihexylfluorene in 17.3 g of toluene was prepared. The poly-

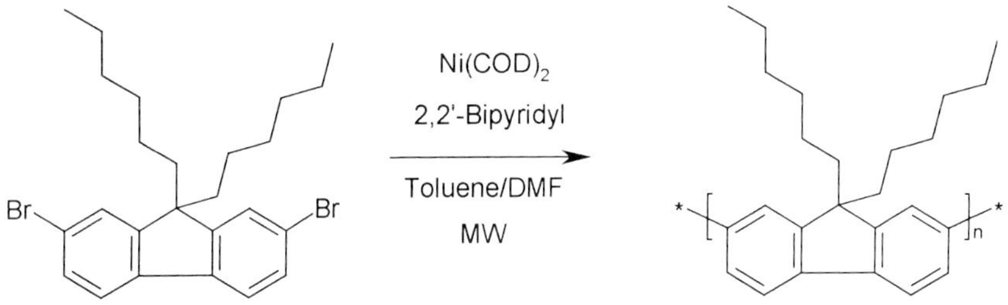

Fig. 3.46 Polymerization of 2,7-dibromo-9,9-dihexylfluorene.

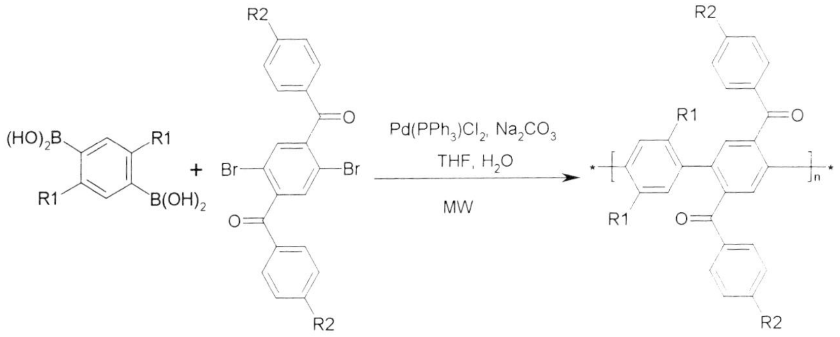

Fig. 3.47 Synthesis of fully conjugated *para*-phenylene ladder polymers.

merization was accomplished by charging a 10-mL reaction vial with 2.2 g of the catalyst solution and 1.77 g of the monomer solution. The vial was sealed and irradiated in a single-mode microwave up to 250 °C for 10 minutes. Eventually, the polymers of number average molecular weight in the range of 5,000 to 10,000 g/mol with PDI between 1.65 to 2.22 were obtained.

The synthesis of fully conjugated *para*-phenylene ladder polymers via microwave-assisted palladium-mediated "double" Suzuki and Still-type reactions was recently demonstrated (Fig. 3.47) (Nehls et al., 2004). The polymerization reactions were run in a THF solution at 130 °C in the presence of palladium catalyst with phosphine ligands and irradiated in a single-mode microwave reactor for 11 minutes. Compared to conventional thermal protocols, the reaction time was reduced from days to a couple of minutes, and molecular weight distributions (PDI approximately 1.8) of the polymers were changed significantly.

Subsequently, thiophene oligomers (up to 6 units) were also obtained

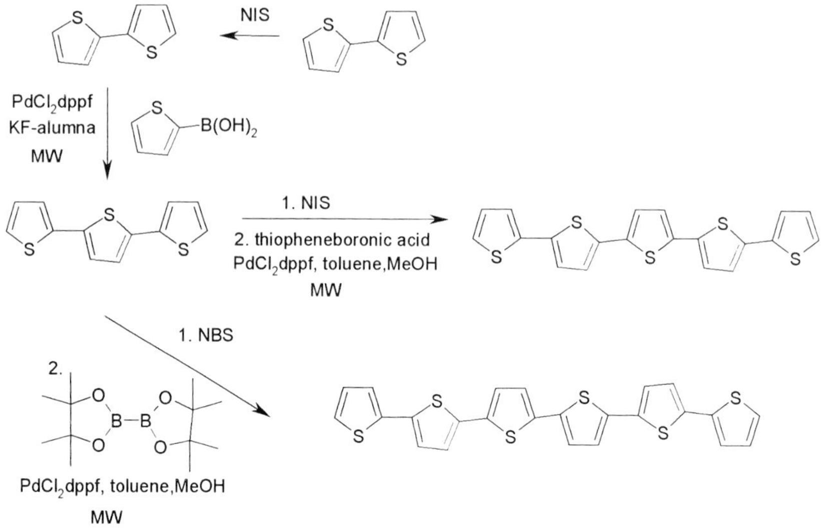

Fig. 3.48   Preparation of thiophene oligomers.

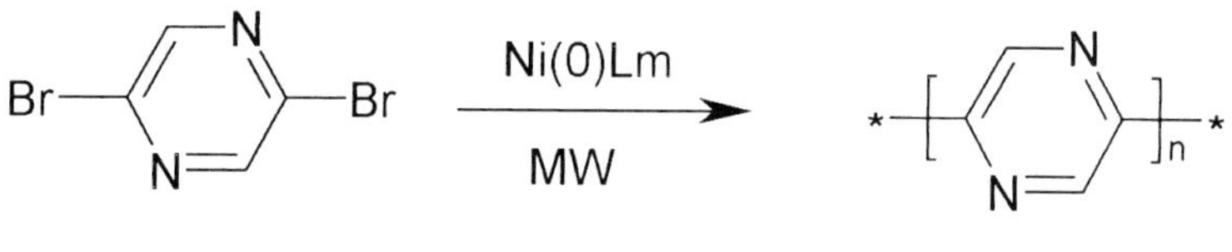

Fig. 3.49   Synthesis of poly(pyrazine-2,5-diyl) polymer.

under microwave conditions from 2-thiophene boronic acid and dibromo precursors with three tiophene units in the presence of a palladium catalyst and KF with KOH (Fig. 3.48). The reactions were run in a single-mode microwave reactor at 70 °C for 10 minutes (Melucci et al., 2004).

Meanwhile, poly(pyrazine-2,5-diyl) polymer was prepared through organometallic dehalogenative polycondensation of 2,5-dibromopyrazine (Fig. 3.49) (Yamamoto et al., 2003). The reaction was carried out by using a mixture of 2.62 mmol of 2,5-dibromopyrazine, 5.23 mmol of bis(1,5-cyclopentadiene)nickel(0), and 5.23 mmol of 2,2'-bipyridyl that was irradiated in a single-mode microwave reactor for 10 minutes in either toluene or DMF solutions. Under microwave conditions, the polymer afforded an 83% to 95% yield, while a similar yield under conventional conditions at 60 °C was obtained after 2 days.

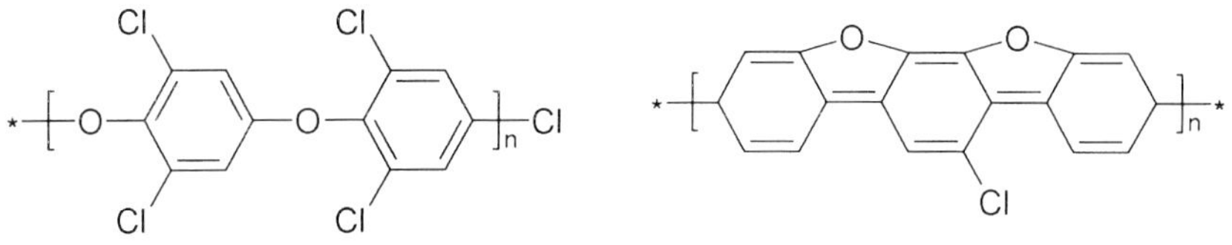

Fig. 3.50 Preparation of poly(dichlorophenylene oxide).

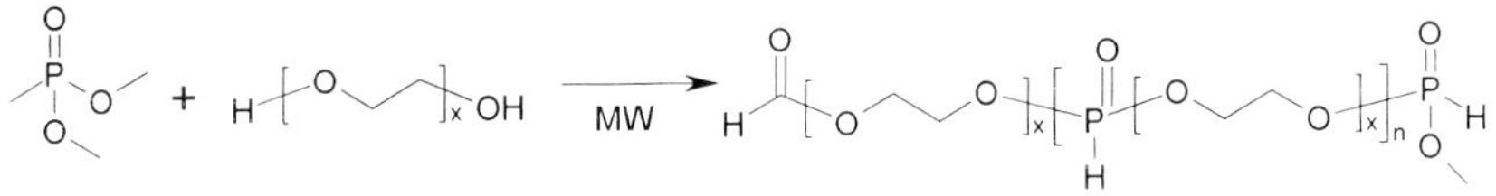

Fig. 3.51 Synthesis of poly(alkylene hydrogen phosphonate)s.

Poly(dichlorophenylene oxide) with a conducting polymer was simultaneously obtained under microwave irradiation from 2,4,6-trichlorophenol (Fig. 3.50) (Cakmak et al., 2004). Microwave-initiated polymerization was performed with a Pyrex vessel in which 2.5 g of 2,4,6-trichlorophenol was mixed with 0.5 g of NaOH and 1 to 2 mL of triple-distilled water. The reaction mixtures were irradiated in a domestic microwave oven for various time intervals of 1 to 7 minutes. The resulting polymers [i.e., poly(dichlorophenylene oxide)] and the conducting polymer (0.3 S/cm^2) were separated by precipitation from toluene. The optimum conditions for poly(dichlorophenylene oxide) and the conducting polymer were 70 W for 5 minutes and 100 W for 1 minute, respectively.

In turn, poly(alkylene hydrogen phosphonate)s were obtained by a transestrification of dimethyl hydrogen phosphate and poly(ethylene glycol) (PEG 400) under microwave irradiation (Fig. 3.51) (Bezdushna et al., 2005). The reactions were carried out in a round-bottomed vessel equipped with an upright condenser in which the mixture of 52.5 mmol of dimethyl hydrogen peroxide and 50 mmol of PEG-400 was irradiated in a single-mode microwave reactor for 55 minutes at 140 to 190 °C. The temperature was monitored by an infrared sensor.

It was found that microwave conditions avoid the undesirable thermal degradation of dimethyl hydrogen phosphate because of short reaction times. Although in comparing microwave and conventional conditions, poly(alkene hydrogen phosphonate)s with higher average molecular weights were obtained under conventional conditions, the reaction time was substantially reduced under microwave conditions from 9 hours to 55 minutes for conventional and microwave conditions, respectively (Table 3.22).

Table 3.22. Comparison of M_n data determined by MALDI-TOF or GPC and by ^{31}P{H} NMR or ^{1}H NMR spectra of the reaction products obtained after heating dimethyl hydrogen phosphonate with poly(ethylene glycol) (PEG400) in a microwave reactor for 55 min. and by conventional heating for 9.5 hours.

| | ^{31}P{H} NMR data | | | ^{1}H NMR data | | | MALDI-TOF[a] or (GPC) |
Method	P intensity in repeating :end groups	n	M_m[b]	P intensity in repeating :end groups	n	M_m[b]	M_n
MW	10.42:1.5	7	3122	1.00:0.14	7	3122	1729
Δ	14.29:1.3	11	4906	1.00:0.09	11	4906	(4700)

[a] The measurement was made in air.
[b] The molecular mass of the repeating unit is 446.
Reprinted from Bezdushna, E., Ritter, H., Troev, K., 2005. Macromol. Rapid Commun. 26:471., with permission.

Last, the modification of polymers under microwave irradiation has been a subject of much research. Because the modification of natural polymers is discussed in Chapter 6, only modification of synthetic polymers is presented in this chapter.

Polyethylene (PE), is probably the most commonly used polymer in everyday life. The modification of PE is one of great importance in the polymer industry (Vasile et al., 2005). The surface oxidation of PE under solid-state conditions under microwave irradiation with potassium permanganate was investigated by Mallakpour and others (2001). Oxidation was performed in a domestic microwave oven, in which the mixture of 2.0 g (0.071 mL) of PE powder ground with 3.76 g (0.024 mol) of KMnO$_4$ was irradiated for 90 seconds. After the reaction, Fourier transform–infrared analysis revealed the presence of hydroxyl as well as vinylic functional groups on the polymer surface, whereas hydroperoxy groups were not detected.

In turn, the modification of oxetane-based polymers with 4-(2-aminoethyl)morpholine (Fig. 3.52) under microwave irradiation was investigated (Hurduc et al., 1996). In a typical experiment, 0.1 g of the polymer with 0.1 to 0.4 g of 4-(2-aminoethyl)morpholine in a DMF solution (10 mL) was exposed to microwave irradiation for 1 to 3 hours in a stereo-mode applicator while the temperature was monitored with use of a thermovision camera.

The highest degree of conversion under microwave irradiation reached 27%, while in several syntheses under conventional conditions, the conversions of oxetane rings were always lower and achieved a maximum value of 18% (Table 3.23).

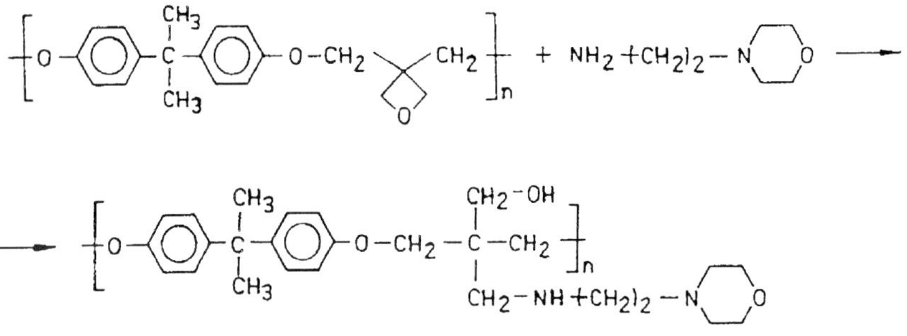

Fig. 3.52 Modification of oxetane-based polymers.

Table 3.23. Effect of various parameters on the degree of substitution in a polymer.

No. sample	Solvent	Temperature (°C)	Time (h)	Degree of substitution (%)
1[a]	$CHCl_3$ + AEM	61	1.0	0
2[a]	$CHCl_3$ + AEM	61	3.0	0
3[a]	DMF + AEM	153	1.0	Traces
4[a]	DMF + AEM	153	2.0	3.5
5[a]	AEM	200	1.0	10.1
6[a]	AEM	200	2.5	27.0
7[b]	$CHCl_3$ + AEM	61	3.0	0
8[b]	DMF + AEM	153	2.0	1.5
9[b]	AEM	205	2.5	18.3

[a] Under microwave irradiation.
[b] Without microwave irradiation.
Reprinted from Hurduc, N., Buisine J.M., Decock, P., Talewee, J., Surpateneau, G. 1996. Polym. J. 28:550, with permission.

References

Danihaohemi, A., Hamamkhani, H., Abidolmaleki, A. 2004. Polym. Sci. Part A Polym. Sci. 42:2106.
Barbbier-Baudry, D., Brachais, L., Cretu, A., Gattin, R., Loupy, A., Stuerga, D. 2003. Environ. Chem. Lett. 1:19.
Bezdushna, E., Ritter, H. 2005. Macromol. Rapid Commun. 26:1087.
Bezdushna, E., Ritter, H., Troev, K., 2005. Macromol. Rapid Commun. 26:471.

Cakmak, O., Bastukmen, M., Kisakerek, D. 2004. Polymer 45:5451.

Caouthar, A., Loupy, A., Bortolussi, M., Blais, J.C., Dubreucq, L., Meddour, A. 2005. J. Polym. Sci. Part A Polym. Chem. 43:6480.

Carter, K.R. 2002. Macromolecules 35:6757.

Chatti, S., Bortolussi, M., Loupy, A., Blais, J.C., Bogdal, D., Majdoub, M. 2002. Eur. Polym. J. 38:1851.

Chatti, S., Bortolussi, M., Loupy, A., Blais, J.C., Bogdal, D., Roger, P. 2003. J. Appl. Polym. Sci. 90:1255.

Chatti, S., Bortolussi, M., Bogdal, D., Blais, J.C., Loupy, A. 2004. Eur. Polym. J. 40:561.

Chen, M., Siochi, E.J., Ward, T.C., McGrath, J.E. 1993. Polym. Eng. Sci. 33:1110.

Chen, Y., Wang, J., Zhang, D., Cai, R., Yu, H., Su, C., Huang, Z.E. 2000. Polymer 41:7877.

Chen, G., Zhu, X., Cheng, Z., Xu, W., Lu, J. 2004. Radiat. Phys. Chem. 69:129.

Cheng, Z., Zhu, X., Zhang, L., Zhou, N., Xue, X. 2003a. Polym. Bull. 49:363.

Cheng, Z., Zhu, X., Chen, M., Chen, J., Zhang, L. 2003b. Polymer 44:2243.

Chia, H.L., Jacob, J., Boey, F.Y.C. 1996. J. Polym. Sci. Part A Polym. Chem. 34:2087.

Chia, L.H.L., Jacob, J., Boey, F.Y.C. 1995. J. Mat. Proc. Techn. 48:445.

Faghihi, K., Hagibeygi, M. 2003. Eur. Polym. J. 39:2307.

Fang, X., Hutcheon, R., Scola, D.A. 2000b. J. Polym. Sci. Part A Polym. Chem. 38:1379.

Fischer, F., Tabib, R., Freitag, R. 2005. Eur. Polym. J. 41:403.

Fleche, G., Huchette, M. 1986. Starch 38:26.

Goretzki, C., Krlej, A., Steffens, C., Ritter, H. 2004. Macromol. Rapid Commun. 25:513.

Hoogenboom, R., Wiesbrock, F., Huang, H., Leenen, M.A.M., Thijs, H.M.L., van Nispen, S.F.G.M., van der Loop, M., Fustin, C.A., Jonas, A.M. Gohy, J.F., Schubert, U.S. 2006. Macromolecules 39:4719.

Hoogenboom, R., Wiesbrock, F., Leenen, M.A.M., Meier, M.A.R., Schubert, U.S. 2005. J. Comb. Chem. 7:10.

Hou, C., Qu, R., Ji, C., Wang, C., Sun, C. 2006. J. Appl. Polym. Sci. 101:1598.

Hurduc, N., Abdelylah, D., Buisine, J.M., Decock, P., Surpateanu, G. 1997. Eur. Polym. J. 33:187.

Hurduc, N., Buisine J.M., Decock, P., Talewee, J., Surpateneau, G. 1996. Polym. J. 28:550.

Imai, Y. 1996a. A new facile and rapid synthesis of polyamides and polyimides by microwave assisted polycondensation. In: Step-Growth Polymers for High Performance Materials: New Synthetic Methods, Hendrick, J.L., Labadie, J.W. (Eds.). pp. 421:430, ACS, Series No 624.

Imai, Y. 1996b. React. Funct. Polym. 30:3.

Imai, Y. 1999. Adv. Polym. Sci. 140:1.

Imai, Y., Nemoto, H., Kakimoto, M. 1996a. J. Polym. Sci. Part A Polym. Chem. 34:701.

Imai, Y., Nemoto, H., Watanabe, S., Kakimoto, M. 1996b. Polymer. J. 28:256.

Iannelli, M., Alupei, V., Ritter, H. 2005. Tetrahedron 61:1509.

Iannelli, M., Ritter, H. 2005. Macromol. Chem. Phys. 206:349.

Jacob, J., Chia, L.H.L., Boey, F.Y.C. 1997. J. Appl. Polym. Sci. 63:787.

Jing, S., Peng, W., Tong, Z., Baoxiu, Z. 2006. J. Appl. Polym. Sci. 100:2244.

Keki, S., Bodnar, I., Borda, J., Deak, G., Zsuga, M. 2001. Macromol. Rapid. Commun. 22:1063.

Leenen, M., Wiesbrock, F., Hoogenboom, R., Schubert, U.S. 2005. e-Polymers 071.

Li, J., Zhu, X., Zhu, J., Cheng, Z. 2006. Radiat. Phys. Chem. 75:253.

Liao, L.Q., Liu, L.J., Zhang, C., He, F., Zhuo, R.X., Wan, K. 2002. J. Polym. Sci. Part A Polym. Chem. 40:1749.

Lu, J., Jiang, Q., Zhu, J., Wang, F. 2001. J. Appl. Polym. Sci. 79:312.

Lu, J., Wu, J., Wang, L., Yao, S. 2004a. J. Appl. Polym. Sci. 97:2072.

Lu, J., Zhu, X., Ji, S., Zhu, J., Chen, Z. 1998. J. Appl. Polym. Sci. 68:1563.

Lu, J., Zhu, X., Zhu, J., Yu, J. 1997. J. Appl. Polym. Sci. 66:129.

Lu, X., Zhu, X., Cheng, Z., Xu, W., Chen, G. 2004b. J. Appl. Polym. Sci. 92:2189.

Mallakpour, S.E., Hajipour, A.R., Mahdavian, A.R., Zadhoush, A., Ali-Hosseini, F. 2001. Eur. Polym. J. 37:1199.

Matyjaszewski, K., Xia, J. 2001. Chem. Rev. 101:2921.

Melucci, M., Barbarella, G., Zambianchi, M., Di Pietro, P., Bongini, A. 2004. J. Org. Chem. 69:4821.

Nehls, B.S., Asawapirom, U., Fuldner, S., Preis, E., Farrell, T., Scherf, U. 2004. Adv. Funct. Mater. 14:352.

Pajda, M., Bogdal, D., Orru, R. 2004. Microwave-Assisted Free-Radical Polymerization Reactions of Various Unsaturated Monomers. In: Modern Polymeric Materials for Environmental Applications, Pielichowski, K. (Ed.). Krakow: 113-116.

Park, K.H., Watanabe, S., Kakimoto, M., Imai, Y. 1993. Polym. J. 25:209.

Palacios, J., Sierra, M., Rodriguez, P. 1992. New Polym. Mater. 3:273.

Pielichowski, J., Bogdal, D., Wolff, E. 2003a. Przem. Chem. 82:8.

Pielichowski, J., Dziki, E., Polaczek, J. 2003b. Pol. J. Chem. Technol. 5:3.

Pielichowski, J., Penczek, P., Bogdal, D., Wolff, E., Gorczyk, J. 2004. Polimery 49:763.

Polaczek, J., Dziki, E., Pielichowski, J. 2003. Polimery 48:61.

Polaczek, J., Pielichowski, J., Pielichowski, K., Tylek, E., Dziki, E. 2005. Polimery 50:812.

Porto, A.F., Sadicoff, B.L., Amorim, M.C.V., Mattos, M.C.S. 2002. Polym. Test. 21:145.

Pourjavadi, A., Zamanalu, M.R., Zohurian-Mehr, M.J. 1999. Angew. Makromol. Chem. 269:54.

Sinnwell, S., Ritter, H. 2005. Macromol. Rapid Commun. 26:160.

Stange, N., Ishaque, M., Niessner, N., Pepers, M., Greiner, A. 2006. Macromol. Rapid Commun. 27:156.

Sundararajan, G., Dhanalakshmi, K. 1997. Polym. Bull. 39:333.

Teffal, M., Gourdenne, A. 1983. Eur. Polym. J. 19:543.

Vasile, C., Pascu, M. 2005. Practical Guide to Polyethylene, Rapra Technology.

Velmati, S., Nagahata, R., Sudiyama, J., Takeuchi, K. 2005. Macromol. Rapid. Commun. 26:1163.

Vogel, B.M., Mallapragada, S.K., Narasimhan, B. 2004. Macromol. Rapid Commun. 25:330.

Watanabe, S., Hayama, K., Park, K.H., Kakimoto, M., Imai, Y. 1993. Macromol. Chem. Rapid. Commun. 14:481.

Wiesbrock, F., Hoogenboom, R., Abeln, C.H., Schubert, U.S. 2004. Macromol. Rapid Commun. 25:1895.

Wiesbrock, F., Hoogenboom, R., Leenen, M.A.M., Meier, M.A.R., Schubert, U.S. 2005a. Macromolecules 38:5025.

Wiesbrock, F., Hoogenboom, R., Van Nispen, S.F.G.M., van Nispen, M., van der Loop, M., Abeln, C.H., van den Berg, A.M.J., Schubert, U.S. 2005b. Macromolecules 38:7957.

Wisnoski, D.D., Leister, W.H., Strauss, K.A., Zhao, Z., Lindsley, C.W. 2003. Tetrahedron Lett. 44:4321.

Wu, H., Li, F., Lin, Y., Yang, M., Chen, W., Cai, R. 2006. J. Appl. Polym. Sci. 99:828.

Xu, W., Zhu, X., Cheng, Z., Chen, G., Lu, J. 2003. Eur. Polym. J. 39:1349.

Yamamoto, T., Fujiwara, Y., Fukumoto, H., Nalamura, Y., Koshihara, S., Ishikawa, T. 2003. Polymer 44:4487.

Yu, Z.J., Liu, L.J. 2004. Eur. Polym. J. 40:2213.

Zhang, C., Liao, L., Liu, L. 2004. Macromol. Rapid Commun. 25:1402.

Zhang, H., Schubert, U.S., 2004. Macromol. Rapid Commun. 25:1225.

Zhu, X., Zhou, N., He, X., Cheng, Z., Lu, J. 2003. J. Appl. Polym. Sci. 88:1787.

Zinck, P., Barbier-Baudry, D., Loupy, A. 2005. Macromol. Rapid Commun. 26:46.

Zong, L. Zhou, S., Sgriccia, N., Hawley, M.C., Kempel, L.C. 2003. J. Microwave. Power. E. E. 38:49.

4

THERMOSETTING RESINS

Microwave-assisted curing of thermosetting polymers was one of the first applications in polymer chemistry and technology and represents the most widely studied areas under both continuous and pulse microwave conditions. The polymer formulations were investigated in terms of structure, dielectric properties, toughness, mechanical strength, percentage of cure, and glass transition temperature. Although most of these applications are in different stages of development and occupy only a minor position on the industrial scale, the situation may change in the near future, particularly for composite materials based on an epoxy matrix.

As thermosetting resins are cured, their dielectric loss factor decreases significantly because of the formulation of cross-linked structures, increasing of viscosity, and decreasing motion of polymer molecules. Thus, thermosetting resins absorb less and less microwaves when they are being cured and the reactions can be self-quenched (Chen et al., 1993). Some examples of these are presented in this chapter.

Epoxy Resins

In some early works, it was found that the pulse method could lead to the fastest heating of the resins (Beldjoudi et al., 1988) and improve their mechanical properties (Thuillier et al., 1989) at the same energy used for the epoxy systems that consist of diglycidyl ether of bisphenol A (DGEBA) with 4,4'-methylenedianiline (DDM) and 4,4'-diaminodiphenyl sulfone (DDS) (Fig. 4.1), respectively. For example, the gel point occurred for DGEBA/DDM system in 19.0 minutes using the pulse length of 0.25 msec and the average power of 40 W, while the gel point was reached in 21.2 minutes for continuous microwave irradiation at 40 W (Beldjoudi et al., 1988).

On the other hand, it was shown that a computer-controlled pulsed microwave processing of epoxy systems that consisted of DGEBA (DER 332)

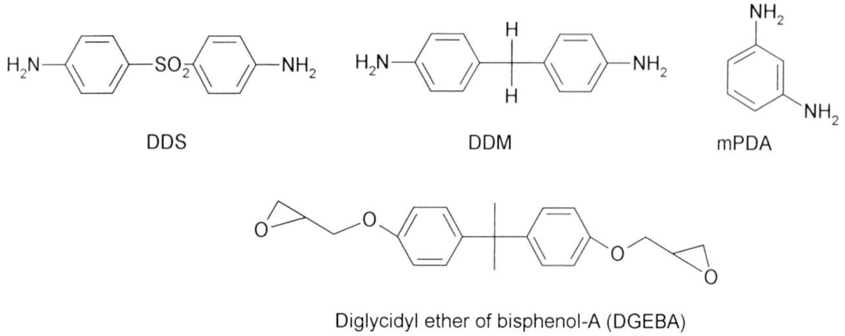

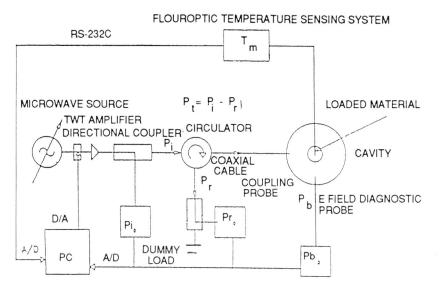

DDS   DDM   mPDA

Diglycidyl ether of bisphenol-A (DGEBA)

Fig. 4.1 Diglycidyl ether of bisphenol A (DGEBA), 4,4'-methylenedianiline (DDM), 4,4'-diaminodiphenyl sulfone (DDS), and *m*-phenylenediamine (mPDA).

and DDS (Fig. 4.1) at a cavity operated in TM_{012} mode could be successfully applied to eliminate the exothermic temperature peak and maintain the same cure temperature to the end of the reaction (Jow et al., 1989). The schematic diagram of the system is presented in Figure 4.2.

It is possible to cure the epoxy systems under pulsed microwave irradiation at higher temperatures and thus it is faster without thermal degradation compared to a continuous microwave and conventional thermal processing.

Fig. 4.2 A computer-aided microwave processing and diagnostic system. Reprinted from Jow, J., DeLong, J.D., Hawley, M.C. 1989. SAMPE Q. 20:46, with permission.

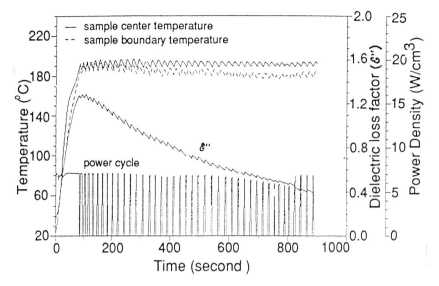

Fig. 4.3 Temperature, loss factor, and input power density measurements during controlled pulsed microwave curing of DER332/DDS resins. Reprinted from Jow, J., DeLong, J.D., Hawley, M.C., 1989. SAMPE Quart 20:46 with permission.

Measurements of dielectric loss factor (ε'') of the samples can also be made during the controlled pulse power cycle, and it can be observed that the loss factor (ε'') decreased with the reaction progress (Fig. 4.3).

Regarding the dielectric behavior of thermosetting systems, the study of the dielectric properties during the curing process has both fundamental and practical applications. Since information about the curing reactions has been provided, nondestructive testing methods to monitor the curing processes have been developed. The dielectric properties of the epoxy system consisted of DGEBA (EPON 828 EL) and ethylenediamine (EDA) was described at the frequency range of 10^3 to 10^{10} Hz (Casalini at al., 1997; Levita et al., 1996). Although the studies were not concerned with microwave processing of epoxy systems, the possibility of using dielectric quantities to obtain information on relevant parameters such as conversion, viscosity change, sol-gel transition, and glass transition temperatures was confirmed. Moreover, the basic parameters such as static dielectric constant (ε_r) and dielectric permeability ($\varepsilon = \varepsilon' - j\,\varepsilon''$) being important for microwave processing can be found in these works. The loss factor (ε'') and permeability (ε') increase with the reaction temperature and decrease with extent of cure, which can be attributed to the higher and lower mobility of molecular dipoles during heating and crosslinking (Fig. 4.4), respectively (Delmotte et al., 1991; Marand et al., 1992).

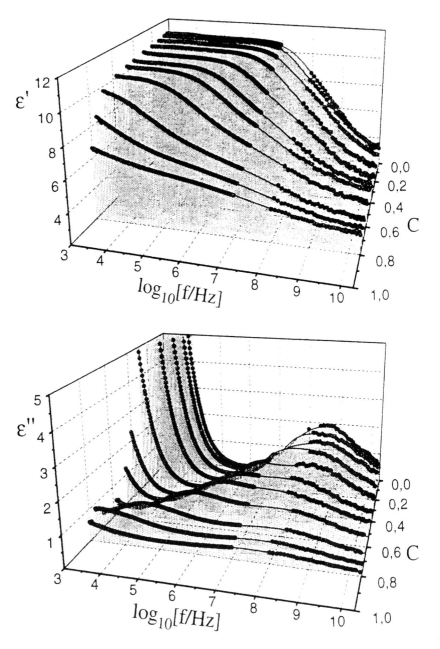

Fig. 4.4 Three-dimensional plot of ε' and ε'' versus conversion C and $\log_{10}$ of frequency for the DGEBA/EDA 1:1 system cured at 25 °C. The black spheres are from experimental data; contour lines of the shaded areas are from fit equations. Reprinted from Casalini, R., Corezzi, S., Livi, A., Levita, G., Rolla, P.A. 1997. J. Appl. Polym. Sci. 65:17, with permission.

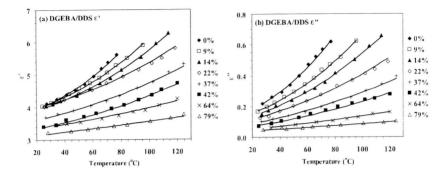

Fig. 4.5 Temperature dependence of ε′ and ε″ at 2.45 GHz for the system of DGEBA and DDS at different extents of cure (%). Reprinted from Zong, L., Kempel L.C., Hawley, M.C. 2005. Polymer. 46:2638, with permission.

Dielectric properties of three curing epoxy resins systems—i.e., DGEBA (DER 332) with DDS, Jeffamine D-230, and *m*-phenylenediamine (mPDA)—at the frequency of 2.45 GHz in the temperature range of 20 to 100 °C were recently investigated (Zong et al., 2005). However, permeability (ε′) and the loss factor (ε″) increased with temperature, which might result in a faster heating rate of the system, they decrease during microwave cure, which is typical for most thermosetting polymers and caused by decreasing number of polar groups in the substrates, increasing of viscosity and decreasing motion of polymer molecules. The theoretical model for calculation of permeability (ε′) and the loss factor (ε″) during curing epoxy systems under microwave irradiation was also elaborated and experimental as well as theoretical data for the DGEBA/DDS system is presented in Figure 4.5 (Zong et al., 2005).

Curing of epoxy systems in thin films was also studied while thermal mechanical analysis (TMA) was used to determine the glass transition temperatures directly from the cured thin film samples (Wei et al., 1993). The epoxy systems consisted of DGEBA with different curing agents (i.e., DDS and mPDA), and the samples were prepared by casting stoichiometric mixtures of DGEBA/DDS and DGEBA/mPDA onto a 13-mm-diameter and 1-mm-thick potassium bromide disks to form approximately 10-μm-thick films that were protected by another potassium bromide discs placed on the top of samples. The thin film samples were cured under microwave irradiation in a center of a cylindrical resonant cavity operating in a TE$_{111}$ mode, while the temperature was monitored by means of fiberoptic thermometer and measured directly from the thin epoxy films. It was found that the effect of microwave irradiation on the cure of epoxy systems depended on the curing agents. Microwaves had a stronger effect on epoxy/DDS sys-

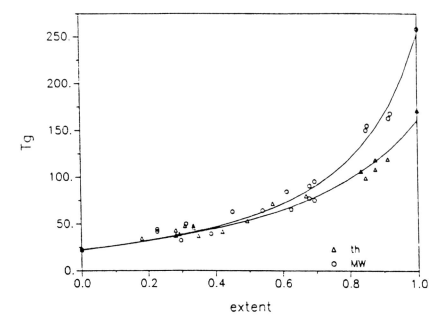

Fig. 4.6   Comparison of glass transition temperatures (T_g) of microwave and thermally cured epoxy resins (DGEBA/DDS). Reprinted from Wei, J., Hawley, M.C., DeLong, J.D., Demeuse, M. 1993. Polym. Eng. Sci. 33:1132, with permission.

tems than on epoxy/mPDA systems, and, consequently, the magnitude of increases in glass transition temperatures was much greater for DGEBA/DDS than for DGEBA/mPDA. Similar values of glass transition temperature were obtained for microwave and thermal cure at low extents of cure, while higher values were observed in microwave cure at the extents of cure after gelation. Moreover, significant higher ultimate extents of cure and faster reaction rates were observed in the microwave cure compared to thermal cure (Figs. 4.6 and 4.7) (Wei et al. 1993).

Consequently, the effect of the microwave curing of the epoxy system consisting of DGEBA and cycloaliphatic diamine (i.e., 4,4'-diamino-3,3'-dimethyldicyclohexyl methane [3DCM]) was also studied for thicker samples and compared with a standard thermal cure (Jordan et al., 1995). DGEBA and the amine (3DCM) were mixed prior to use, and then the epoxy mixture (13 g) was poured into molds (inside dimension 96 mm × 16 mm × 8 mm), which were irradiated in a microwave applicator with TF_{01} propagation mode. The sample temperature was measured continuously by means of an infrared pyrometer that gave the surface temperature and a fiberoptic thermometer that recorded the bulk temperature. Samples

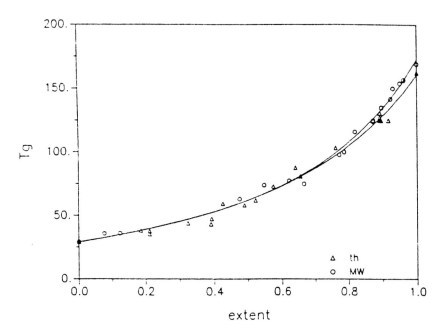

Fig. 4.7 Comparison of glass transition temperatures (T$_g$) of microwave and thermally cured epoxy resins (DGEBA/mPDA). Reprinted from Wei, J., Hawley, M.C., DeLong, J.D., Demeuse, M. 1993. Polym. Eng. Sci. 33:1132, with permission.

cured by both thermal and microwave processing were characterized by dynamic and static mechanical properties and then compared with those of fully crosslinked networks. It was found that microwaves did not have direct influence on the mechanical processing of the polymer network, and the only parameter that influenced the mechanical properties was the extent of the reaction. Moreover, under microwave processing conditions, it was not possible to obtain a fully cured DGEBA/3DCM network. The fully cured samples were obtained either by thermal (140 °C, 1 hour + 190 °C, 14 hours) or combined processing (microwave 200 W, 15 minutes + thermal 190 °C, 14 hours) (Table 4.1) (Jordan et al., 1995).

Later, the kinetic study of a neat and rubber-modified epoxy formulation cured under microwave and thermal conditions was conducted to compare the behavior of the epoxy-amine system during microwave and conventional conditions (Hedreul at al., 1998). The epoxy formulation consisted of DGEBA and DDS as a curing agent with the epoxy-terminated butadiene-acrylonitrile random copolymer (ETBN), which was prepared in the reaction of CTBN with an excess of DGEBA in the presence of triphenylphosphine as a catalyst (Verchere et al., 1990). The weight percentage of rubber

Table 4.1.   The comparison of epoxy compositions cured under both microwave and microwave/conventional conditions.

Curing cycle	MW	MW + Δ
Power (W)	200	200
Temperature (°C)		190
Time	15 min	15 min + 14 h
Compression modulus (GPa)	3.15	2.9
Poisson's ratio	0.36	0.36
Glass transition temperature (°C)	131	186
Extent of reaction (%)	89	100

Reprinted from Jordan, C., Galy, J., Pascault, J.P., More, C., Delmotte, M., Jullien, H. 1995. Polym. Eng. Sci. 35:233, with permission.

introduced into the modified epoxy component amounts to 15% in respect to the total amount of the mixture. Under microwave conditions, the samples (12.5 g) were cured in a microwave applicator with the propagation mode TE_{01} at 170 to 190 °C for 0 to 3 hours. The same kinetic behavior of the epoxy-amine DGEBA/DDS and the ETBN-modified systems was observed whichever curing process (conventional or microwave) was employed. The determined conversion values of the samples cured by microwave irradiation were compared with the kinetic model, which was derived during this study from the conventional thermal curing, and found to be independent of the curing method (Fig. 4.8) (Hedreul at al., 1998).

In situ real-time studies on the mechanism and rate of reaction for epoxy systems under thermal and microwave conditions were also conducted for a number of nonpolymer-forming and polymer-forming mixtures of different functionality, and molecular architecture of the resulting polymers was established as well (Mijovic et al., 1998). In the case of epoxy systems, the reaction of phenylglycidyl ether (PGE) and aniline as a control reference was investigated, whereas for epoxy crosslinking systems, the reactions of diglycidyl ether of bisphenol F (DGEBF) with DDS and DDM were chosen (Fig. 4.9).

In this case, the samples of reaction mixtures were held in a cylindrical sample holder (10-mm diameter and 25-mm depth), which was placed in a waveguide operated under TE_{01} mode for the microwave experiments. The sample temperature during cure was monitored with a fiberoptic thermometer by placing the fiberoptic probe in the center of the sample, while the remote receiving and transmission legs of optical fibers for near-infrared spectrometer were inserted through the holes drilled on each side (Fig. 4.10) (Mijovic et al., 1998).

The temperature range investigated in this study was between 70 and 120 °C and 140 and 190 °C for DGEBF/MDA and DGEBF/DDS, respectively.

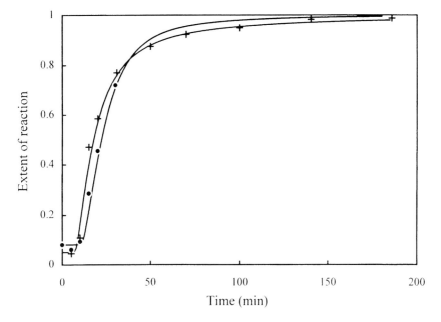

Fig. 4.8 Conversion versus time for DGEBA-DDS (*crosses*) and DGEBA-ETBN (*filled circles*) samples cured by microwave heating; lines represent the models prediction. Reprinted from Hedreul, C., Galy, J., Dupuy, J., Delmotte, M., More, C. 1998. J. Appl. Polym. Sci. 68:543, with permission.

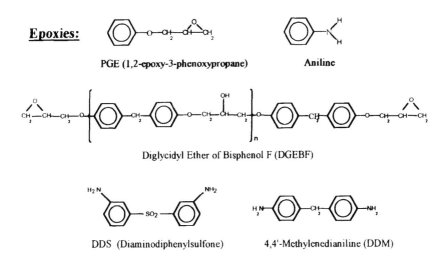

Fig. 4.9 Chemical structures of formulation components for curing epoxy resins.

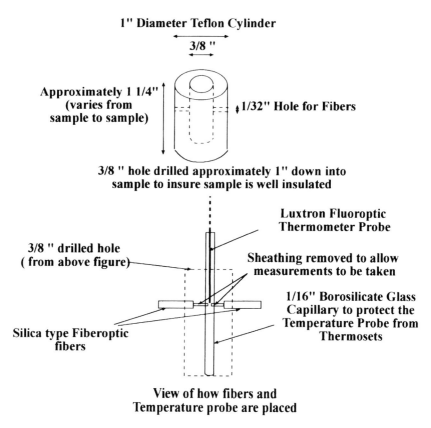

Fig. 4.10   Schematic of a sample cell configuration. Reprinted from Mijovic, J., Corso, W.V., Nicolais, L., d'Ambrosio, G. 1998. Polym. Adv. Technol. 9:231, with permission.

A program was implemented to control the temperature by controlling the power to the microwave generator. On average, the power was on for more than 80% of time. The configuration of the entire microwave setup is shown schematically in Figure 4.11.

Similar to the previous report on the rate and mechanisms of epoxy-amine reactions in thermal and microwave fields (Mijovic et al., 1992), it was found that the cure kinetics of the multifunctional epoxy formulation are identical under both thermal and microwave conditions (Figs. 4.12 and 4.13). The reproducibility and reliability of data were verified by repeated runs, and the claims of accelerated kinetics due to the so-called microwave effect were unfounded; however, an important scientific research issue remains with heat transfer in thermal and microwave fields.

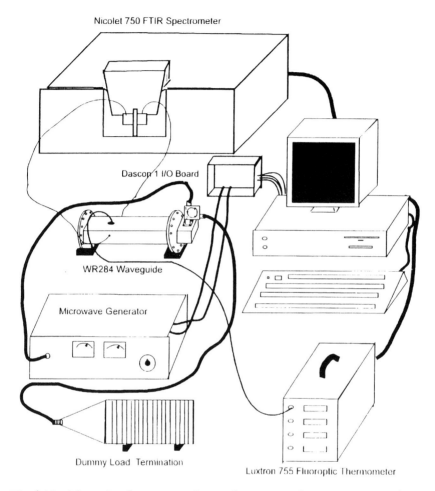

Fig. 4.11 Schematic of experimental setup for in situ real-time monitoring of microwave processing by remote fiberoptic near-infrared spectroscopy. Reprinted from Mijovic, J., Corso, W.V., Nicolais, L., d'Ambrosio, G. 1998. Polym. Adv. Technol. 9:231, with permission.

On the other hand, in a recent paper in which microwave irradiation and thermal curing were performed on the epoxy systems, the reaction rate enhancement under microwave conditions was reported (Boey et at., 1999). The three types of curing agents were used (i.e., DDS, DDM, and mPDA with DGEBA [Araldite GY6010]). Microwave curing of epoxy systems was carried out in a multimode cavity (300 mm × 298 mm × 202 mm) coupled through a waveguide with a variable-power DC generator up to 1.26 kW. A thermocouple covered in a stainless steel sheath was rigidly mounted above the center of the cavity and electronically connected to the cavity so

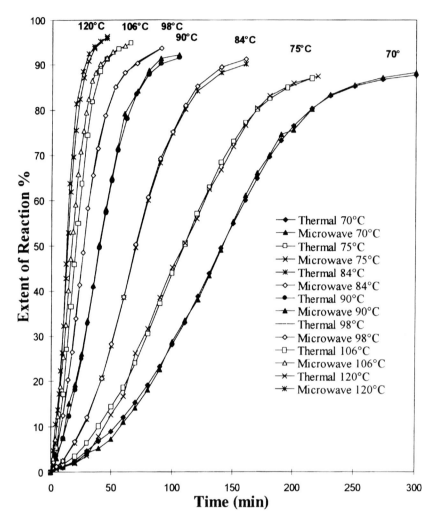

Fig. 4.12   Extent of reaction as a function of time in thermal and microwave fields for DGEBF-MDA reaction at different temperatures. Reprinted from Mijovic, J., Corso, W.V., Nicolais, L., d'Ambrosio, G. 1998. Polym. Adv. Technol. 9:231, with permission.

that microwaves would not penetrate into the sheath. After the radiation, the samples were removed with the specimen temperature (T_{spec}) measured by the thermocouple and cooled and stored in a freezer. Finally, the percentage cure of the resin mixtures at selected time intervals was calculated from the heat of reaction of the exotherm curve obtained during the DSC run of the samples.

The results indicated that in all three cases, the curing time was consi-

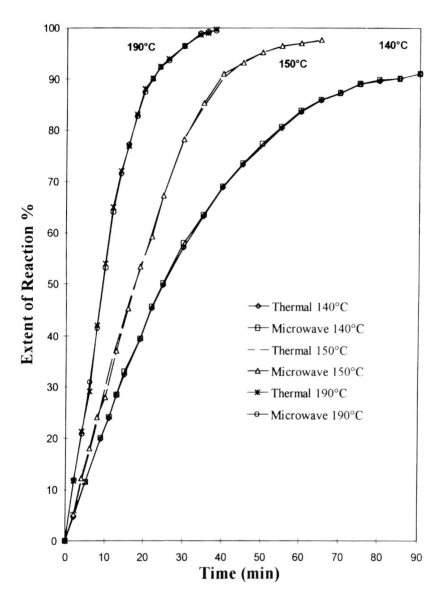

Fig. 4.13 Extent of reaction as a function of time in thermal and microwave fields for DGEBF-DDS reaction at different temperatures. Reprinted from Mijovic, J., Corso, W.V., Nicolais, L., d'Ambrosio, G. 1998. Polym. Adv. Technol. 9:231, with permission.

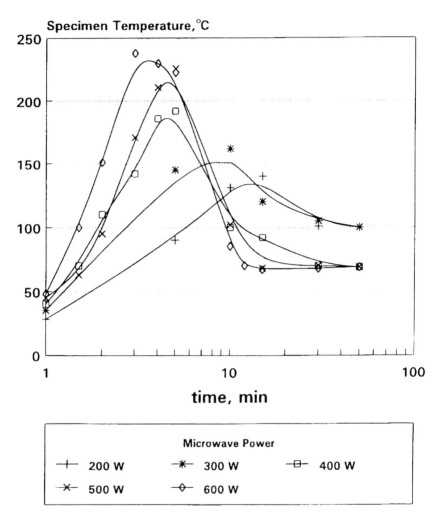

Fig 4.14 Plot of the specimen temperatures versus cure time during isothermal curing for epoxy-DDS system. Reprinted from Boey, F.Y.C., Yap, B.H., Chia, L. 1999. Polym. Test. 18:93, with permission.

derably shortened. In contrast to thermal curing, where the DDS system took a longer time to fully cure, for microwave curing at higher power levels, the differences in time to fully cure for all three systems were minimal. However, for the DDM and mPDA systems, full curing was achieved at all power levels; the percentage cure for the DDS system could not reach 100% at a lower power level. For instance, under microwave conditions, at a low power level (i.e., 200 W), the DDM and mPDA systems appeared to give shorter curing times of 10 to 8 minutes, respectively, in comparison

Table 4.2. Maximum of specimen temperature (T_{spec}) of microwave-treated epoxy/amine systems.

Microwave curing temperature (°C)	T_{spec} (°C)	Cure time (min)
DGEBA/DDS		
200	141	15
300	165	10
400	184	5
500	225	5
600	235	3
DGEBA/DDM		
200	111	10
400	137	5
600	174	3
DGEBA/mPDA		
200	135	8
400	150	4
600	212	2.5

Reprinted from Boey, F.Y.C., Yap, B.H., Chia, L. 1999. Polym. Test. 18:93, with permission.

with the DDS system, which cured after 15 minutes. However, at higher power levels (i.e., 400 and 600 W), the curing time was almost the same: 2.5 minutes for mPDA and 3 minutes for both DDS and DDM systems (Fig. 4.14 and Table 4.2) (Boey et at., 1999).

According to the next report, in which the same experimental methodology was applied, the observed rate enhancement was the result of the decrease in the lag time prior to incitation of crosslinking and, in consequence, the decrease in the overall effective cure time (Boey et al., 2000). Because a shortening of cure time is no different from a shortening achieved by a higher curing temperature, by comparison of the curing processes under both microwave and thermal conditions, it was possible to obtain temperature equivalent values for the microwave cure, which are only virtual ones and are prepared for the purpose of analysis of the cure kinetics. The regression plot of the effective cure time for thermal curing to obtain the equivalent temperature for microwave curing and equivalent temperature for microwave curing are presented in Figure 4.15 and Table 4.3, respectively.

It is worth noting that the equivalent temperatures obtained in all the cases were consistent and significantly higher than the maximum temperature measured in the samples, providing further support that microwave curing is not merely thermal based (Boey et al., 2000).

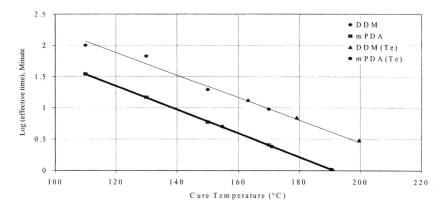

Fig. 4.15   Regression plot of the effective cure time for thermal curing to obtain the equivalent temperatures for microwave curing. Reprinted from Boey, F.Y.C., Rath, S.K. 2000. Adv. Polym. Techn. 19:194, with permission.

More recently, microwave curing processes of the epoxy systems were investigated in order to observe the effect of different curing agents and time of curing in the final cured glass-transition temperature (T_g) (Boey et al., 2001). Microwave irradiation and thermal curing were performed on DGEBA with three curing agents: DDS, DDM, and mPDA. Microwave curing was conducted in a multimode cavity that was again powered with a variable-power generator up to 1.26 kW at 2.45 GHz. All three systems (i.e., DDS, DDM, and mPDA) exhibited a shorter curing time to reach the maximum percentage cure and T_g; the actual maximum values achieved for

Table 4.3.   Equivalent cure temperature for microwave processing of epoxy/ amine systems.

Microwave power (W)	Maximum temperature (°C)	Lag time (s)	Effective cure time (s)	Total cure time (s)
	DGEBA/DDM system			
200	120	144	163	154
400	130	170	179	173
600	162	181	196	190
	DGEBA/mPDA system			
200	135	150	153	152
400	151	171	172	170
600	178	193	190	191

Reprinted from Boey, F.Y.C., Rath, S.K. 2000. Adv. Polym. Techn. 19:194, with permission.

Table 4.4. Time to maximum percentage cure and T_g for thermal and microwave curing.

	DEBGA/DDS (min)		DEBGA/DDM (min)		DEBGA/ mPDA (min)	
	Percentage cure	T_g	Percentage cure	T_g	Percentage cure	T_g
Cure temp. (°C)			Thermal curing			
110	420	640	10	34	27	28
130	185	550	8.5	16	10	8
150	115	280	7	7	4	4
170	70	188				
180	32	150				
Cure power (W)			Microwave curing			
200	54	53	17	20	7	9
400	7	7	6	6	2.5	2.5
600	1.5	2	3.5	3.5	1	1

Reprinted from Boey, F.Y.C., Yap, B.H. 2001. Polym. Test. 20:837, with permission.

both the maximum percentage cure and T_g were significantly lower than those for thermal curing. While a faster a rate was obtained during microwave curing, the existence of highly electron attracting groups (i.e., -SO_2- [DDS]) appeared to induce a delay in the reactivity of the amine functions and effectively inhibit further curing. In contrast, both DDM and mPDA systems were able to achieve the maximum percentage cure and T_g values equal to those for thermal curing at a significantly shorter curing time due to a greater reduction in the effective cure time than in the lag time. Microwave curing was more effective in reducing the overall cure time for mPDA system, which suggested that the microwave curing was more effective in enhancing the reaction rates during crosslinking rather than shortening only the lag time (Table 4.4) (Boey et al., 2001).

A similar investigation into microwave-assisted curing of DGEBA with maleic anhydride showed that compared to conventional thermal cure, microwave cure can decrease the curing temperature by 15 to 20 °C, increase the compressive strength and bending strength of the epoxy resin, and reduce the amount of maleic anhydride by amount of 5% (Table 4.5) (Zhou et al., 2003).

In turn, the microwave-assisted method of synthesis of high-molecular-weight (solid) epoxy resins was described (Bogdal et al., 2003, 2004). The

Table 4.5. Mechanical properties of epoxy resin cured by microwave and thermal treatment.

Curing process	Maleic anhydride Phr	Curing time (h)	Curing temperature (°C)	Compressive strength (MPa)	Bending strength (MPa)
Microwave	25	1	135 + 5	91.01	35.60
Microwave	30	1	135 + 5	105.07	56.68
Microwave	35	1	135 + 5	92.29	44.84
Microwave	40	1	135 + 5	93.97	36.81
Thermal	30	3	150 + 5	88.68	—
Thermal	35	3	150 + 5	90.38	34.76
Thermal	40	3	150 + 5	88.54	—

Reprinted from Zhou, J., Shi, C., Mei, B., Yuan, R., Fu, Z. 2003. J. Mater. Process. Technol. 137:156, with permission.

method is based on the polyaddition reaction of bisphenol A to a low-molecular-weight epoxy resin or DGEBA in the presence of 2-methylimidazole (2-MI), as a catalyst (Fig. 4.16). The syntheses were also performed using conventional thermal heating (i.e., electric heating mantle) for the comparison of properties of high-molecular-weight epoxy resins under both microwave and conventional conditions.

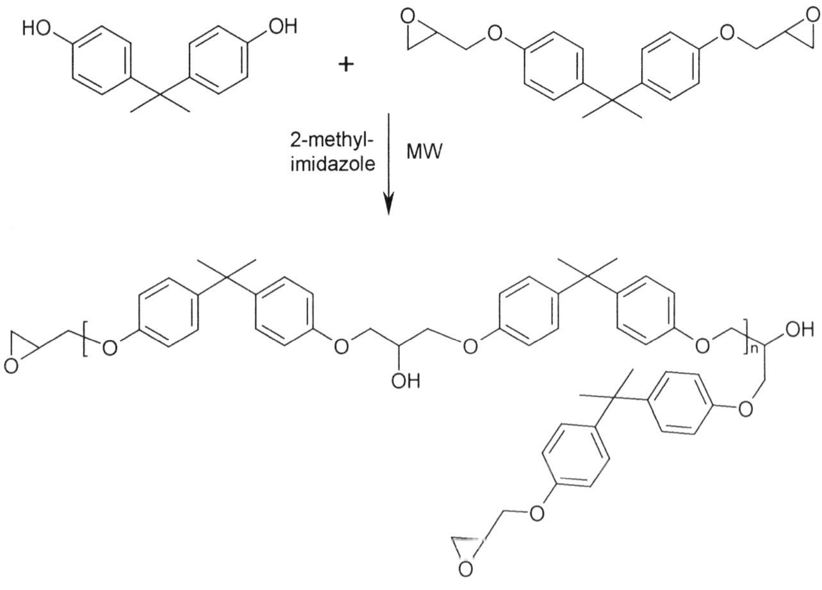

Fig. 4.16 Synthesis of high-molecular-weight (solid) epoxy resins.

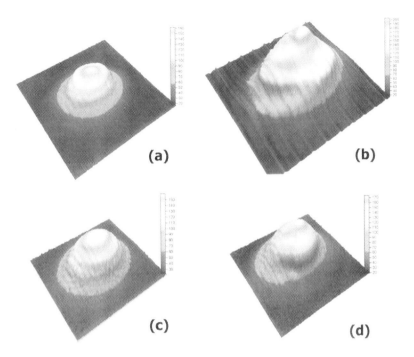

Fig. 4.17 Surface temperature of the solid epoxy resins samples monitored by means of the thermovision camera.

of epoxy resins with bisphenol A, in some parts of a sample local temperature under microwave irradiation can be higher than the bulk temperature of the reaction mixture and, therefore, even pyrometers as well as fiberoptic thermometers are not capable of giving correct information about temperature. Moreover, in case of a high temperature gradient of processed material that can be observed with the maximum temperature at the center of sample, it is hard to state which is the bulk temperature of the reaction mixture (Bogdal et al., 2004a).

Recently, the same approach for the synthesis of solid epoxy resins with reduced flammability was presented (Brzozowski et al., 2006). For this purpose, bisphenol A was either substituted or partially substituted with 1,1-dichloro-2,2-bis(4-hydroxyphenyl)ethylene (Fig. 4.18), and the synthesis of solid epoxy resins was realized in the same manner as described previously.

Polymerization of some other epoxides such as 3,4-epoxycyclohexylmethyl or 3,4-epoxycyclohexylcarboxylate initiated by diaryliodonium or triarylsulfonium salts (Fig. 4.19) under microwave conditions was studied as well (Zhang at al., 2004). The syntheses were carried out in a multimode mi-

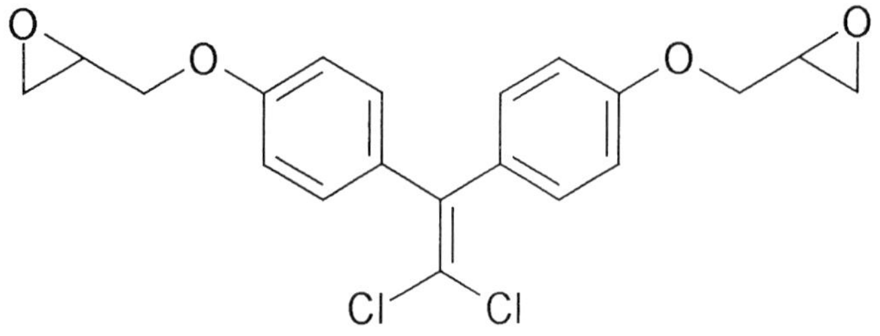

Fig. 4.18   1,1-Dichloro-2,2-bis(4-hydroxyphenyl)ethylene.

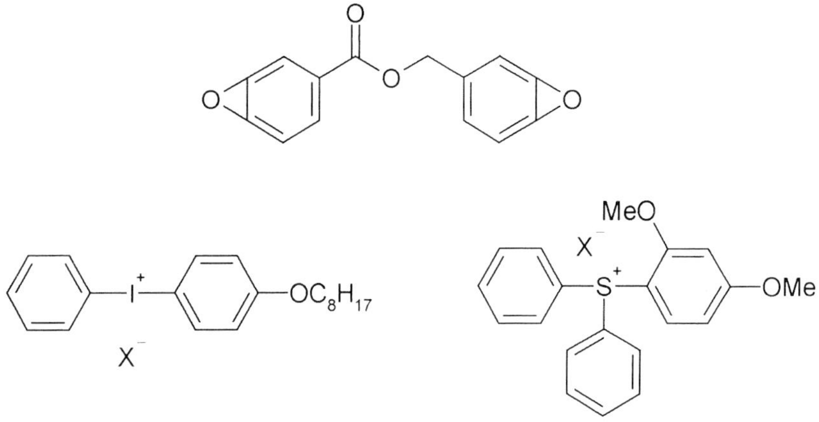

Fig. 4.19   Polymerization of 3,4-epoxycyclohexylmethyl and 3,4-epoxycyclohexylcarboxylate initiated by diaryliodonium or triarylsulfonium salts.

crowave reactor in which 20-mL vials with 4-g samples were placed and irradiated. However, the temperature for microwave polymerizations was measured with the thermocouple immediately after it was removed from the microwave reactor; the extent of the polymerization was determined by means of DSC and Fourier transform-infrared and compared with samples cured under conventional conditions.

The performed study revealed some polymerization phenomena such as polymerization selectivity, polymerization temperature shift, and polymerization temperature shift by microwave power setting under microwave irradiation in comparison with conventional conditions. To explain the phenomena, it was proposed that a new dipole partition function exists in the microwave field so the values of thermodynamic properties such as inter-

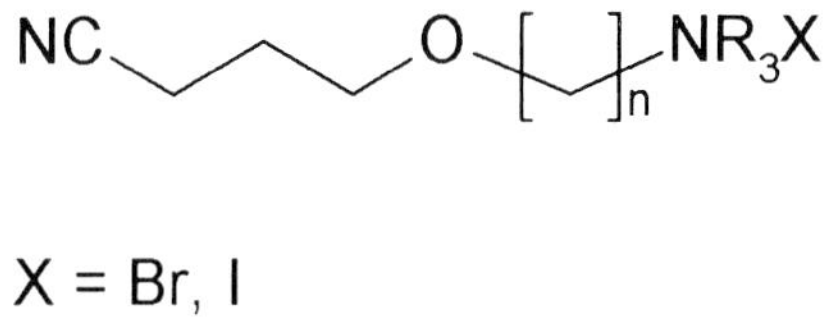

NC ⌒⌒ O ⌊ ⌋ₙ NR₃X

X = Br, I

Fig. 4.20 *N*-(Cyanoalkoxyalkyl)trialkylammonium halides.

nal energy and Gibbs free energy of the material with permanent dipole moments changed under microwave conditions, which in turn led to shifts in the reaction equilibrium and kinetics compared to conventional conditions at the same temperature (Zhang at al., 2004b).

The application of microwave irradiation to cure isocyanate/epoxy resins in the presence of *N*-(2-hydroxyalkyl)trialkylammonium halides was also claimed to impart accelerations to both curing and postcuring kinetics with respect to conventional hot-air heating (Parodi et al., 1994). As a consequence of this investigation, a new class of catalysts that endow aromatic isocyanate/epoxy and aliphatic or cycloaliphatic epoxy/anhydride systems with a particular efficiency for microwave processibility was developed (Parodi et al., 1996). The catalysts belong to the family of *N*-(cyanoalkoxyalkyl)trialkylammonium halides (Fig. 4.20), and their evaluation of the microwave enhancements was performed via isothermal comparative curing experiments under hot-air and microwave heating (Parodi, 1999).

All the curing experiments of isocyanate/epoxy resins were performed on 3-g samples of various liquid resins that were placed in 5-mL PTFE vials. Microwave irradiation was carried out in a cylindrical (diameter 18.5 cm) single-mode tunable microwave cavity. In both the thermal and microwave experiments, the sample temperature was monitored by a fiberoptic probe that was immersed at the center of samples (Table 4.7). As a result, the strong reaction enhancements of the specific catalysts were imparted under microwave heating to all of reactive systems examined. The gelation and vitrification times were reduced by a factor of 8 to 10 of those under hot-air heating with the same catalyst and its concentration. Finally, an ion-hopping conduction mechanism has been recognized as the dominant source of the microwave absorption capacities of these catalysts (Parodi, 1999).

Polyurethanes

The high versatility of the polyurethane raw materials results in the manufacturing of a large variety of products with diverse structures and polymer

Table 4.7. Isothermal curing times of aromatic isocyanate/epoxy and aliphatic epoxy/anhydride resin systems under conventional and microwave heating.

Resin system curing type (temperature)		Catalyst and concentration (mmol/100 g)	Curing time	
			Gelation	Vitrification
System A: l-MDI/DGEBA				
70:30 w/w		I-1[a]	40 min	>80 min
Conventional	(66 °C)	1.90	?	15–20 min
Microwave	(66 °C)	1.90		
System B: ERL-4299/				
(SA/Me-HHPA) 1:1 w/w		I-2[b]	420 min	870 min
Conventional	(90 °C)	1.08	50 min	110 min
Microwave	(90 °C)	1.08		
System C: ERL-4234/DDSA		I-3[c]		
Conventional	(70 °C)	1.09	25.5 h	63 h
Microwave	(70 °C)	1.09	3 h	6 h
System C: ERL-4234/DDSA		I-4[d]		
Conventional	(85 °C)	1.14	270 min	450 min
Microwave	(85 °C)	1.14	30 min	60 min
		TBAI[e]		
Microwave	(85 °C)	1.52	210 min	?

[a] N-[3-(2-Cyanoethoxy)propyl]-N,N-dimethyldecylammonium iodide.
[b] 4-[3-(2-Cyano-ethoxy)ethyl]-4-butylmorpholinium iodide.
[c] N-[2-(2-Cyanoethoxy)propyl]-N,N-dimethyl-decylammonium iodide.
[d] 4-[3-(2-Cyanoethoxy)ethyl]-4-butylmorpholinium iodide.
[e] Tetrabutylammonium iodide.

Reprinted from Parodi et al. 1999. Microwave heating and the acceleration of polymerization processes. In: Polymers and Liquid Crystals, Wlochowicz (ed.) Proceedings of SPIE—The International Society for Optical Engineering, 4017:2, with permission.

matrix from flexible to rigid, especially cellular materials. In the middle of the 1980s, one of the studies of crosslinking polyurethane resins under microwave irradiation was realized by means of pulsed microwave irradiation (Jullien et al., 1985). Thermal behavior of a polyisocyanate-polyalcohol mixture (i.e., 75% ethyl acetate solution of two prepolymers: triisocyanate [Desmodur L75] and polyester-polyalcohol [Desmophen 800]) taken in stoichiometric amounts was investigated, as well as the formation of polyurethane coatings from the same mixture with film hardness as a function of different pulse regimens. In order to cancel the substrate effect, the prepolymer solution (1.4 g) was poured in hollowed PTFE plates, which were the same size and had the same amount of solution, so that the microwave cavity filling factor only varied as a function of the complex dielectric constant in the evolved material. The experimental set-up included a microwave generator (2.45 GHz, 3 kW), provided with an auxiliary pulse generator and a DC generator for power level adjustment. During the polyurethane resin formation, the microwave average power was 30 W and the peak pulse power varied from 200 to 2500 W; the pulse period was from 2 to 200 msec, so the pulse time was varied from 50 μsec to 30 μsec (Fig. 4.21).

It was found that the energy transfer by pulsed microwaves is more efficient than that by an energy equivalent continuous wave value. For instance, in Figure 4.22, the variations of the maximum temperature (T_{max}) with peak power are reported where each curve is related to a pulse period. All curves are issued from the same point (on the left side), corresponding to the reference continuous wave (i.e., power 30 W, T_{max} 70 °C). Starting from this point, T_{max} always increases with pulse power (Fig. 4.22). Thus, it can already be concluded that the energy transfer by pulse microwaves is more efficient than that by continuous microwave irradiation.

Finally, it was found that thermal kinetics, film hardness, and thus macromolecular networks were influenced by microwave pulse irradiation, so that microwave-cured polyurethane films were very much harder than oven-cured materials (Jullien et al., 1985).

The effect of microwave irradiation on the crosslinking of polyurethane resin consisted of two components—diisocyanate derived from 4,4′-diisocyanate diphenylmethane and a low viscosity polyethertriol was then investigated (Silinski et al., 1987). The crosslinking was carried out without a catalyst, while all the samples of polyurethane resin were crosslinked inside a waveguide with TE_{01} propagation mode. The curing of the polyurethane resin samples irradiated at power values from 10 to 60 W led to final networks with mechanical properties of quality at least equivalent to those prepared under conventional thermal conditions. For example, the average elasticity modulus determined from unaxial compression with samples (25-mm height and 12.5-mm diameter) was equal to 3,120 MPa for curing under microwave conditions for 1 hour at 20 W and 2,810 MPa for thermal curing in an oven at 60 °C for 8 hours (Silinski et al., 1987).

Rigid polyurethane foams are widely used as thermal insulating materials. The high versatility of the polyurethane raw materials allow manufacturers to produce a large variety of products with diverse structures and properties. Recent trends have driven the industry to apply environmentally benign renewable components that are less tolerant of factors such as odor and contamination (Prociak et al., 2004; Randall et al., 2002). The application of microwave irradiation is expected to be helpful in the chemical processes of the preparation of novel raw materials for polyurethane foams.

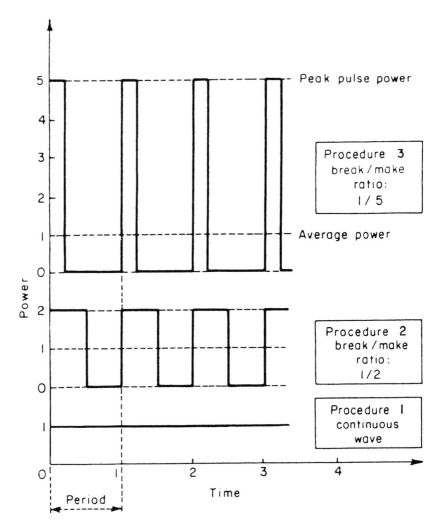

Fig. 4.21 An example of three pulse procedures with the same average micro-wave power transferred to materials. Reprinted from Jullien, H., Valot, H. 1985. Polymer 26:506, with permission.

The vegetable oil–based polyols are examples of such components that can be used to obtain rigid polyurethane foams that have satisfied use proper-ties. For instance, a two-step process was applied for the preparation of rapeseed oil– and linseed oil–based polyols (Prociak et al., 2006a). In the first step, unsaturated fatty acids of triglycerides reacted with acetate per-oxyacid to form epoxidized oil, in which the double bonds of the trigly-

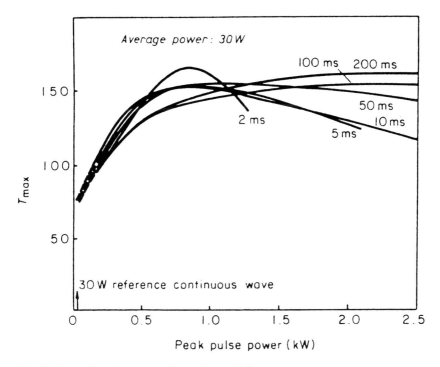

Fig. 4.22 Peak pulse power dependence of the maximum temperature T_{max} (second parameter: pulse period). Reprinted from Jullien, H., Valot, H. 1985. Polymer 26:506, with permission.

cerides were transformed into oxirane rings. In the second step, the epoxidized oils were converted into the polyols in the reaction of the epoxy rings with monoethylene glycol (MEG) or diethylene glycol (DEG). In the case of the polyol based on rapeseed and linseed oils, it was reported that microwave irradiation can be applied for both steps of the process (i.e., the epoxidation and oxirane ring opening reactions). It resulted in a reduction in the reaction time of the epoxidation reaction approximately 60%, and, then in the hydroxylation step approximately 75% in comparison to the processes under conventional conditions. Further comments related to this topic are given in Chapter 5.

The influence of microwave irradiation on blowing and curing processes of rigid polyurethane foams was also studied (Prociak et. al., 2006b). For this purpose, the polyurethane foams were blown inside two different microwave cavities with microwave power regulation by a turn-off and turn-on duty cycle (periodical proportioning) and with continuous microwave power regulation. The foams were synthesized in one-step processes; component B (isocyanate) was added to component A (polyol premix with additives) and the mixtures were stirred for 10 seconds with an overhead stirrer. Then the prepared mixtures were dropped into a mold. The foaming processes were realized as a free rise in the open mold. The reaction mixture was expanded freely in the vertical direction on the square area of 20 cm × 20 cm in polypropylene molds at ambient temperature of approximately 20 °C. In the second case, the polypropylene molds were placed inside a microwave reactor with a continuous power regulation at the range of 30 to 600 W.

Under the conditions of periodical proportioning of microwave irradiation, the expanding processes were not well settled, and the phenomena of the rise and collapse of the foams were observed, which in comparison to the products obtained under continuous microwave irradiation resulted in worse properties and cell structure of these foams (i.e., heat-insulating properties and increase of open cell content, respectively). Therefore, further investigations were carried out only in the microwave reactor with continuous power regulation. Then the tests of polyurethane foaming in the microwave reactor with the continuous power regulation were focused on the proper choice of microwave power and irradiation time because these parameters significantly influence the blowing process and affect the mechanical and physical properties of the rigid polyurethane foams. The results presented in Figures 4.23 and 4.24 reflect the important dependence

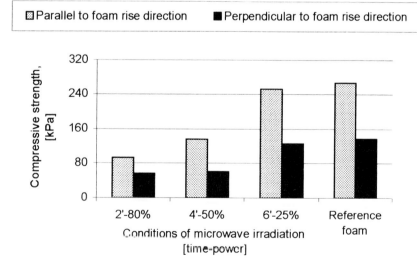

Fig. 4.23 Compressive strength of polyurethane foams measured in parallel and perpendicular rise directions.

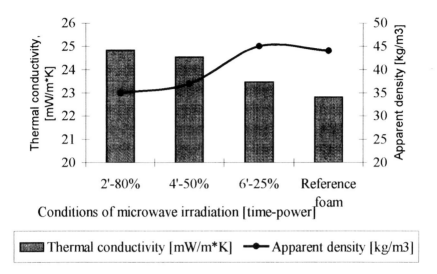

Fig. 4.24 Thermal conductivity and apparent density of polyurethane foams.

of foam properties on such parameters of microwave irradiation as power and time. Application of lower power during a longer time resulted in the rigid foams with more beneficial properties.

In general, the growth of apparent density of rigid polyurethane foams determines their better mechanical properties. However, in this case, it must be stressed that an increase of apparent density by approximately 20% brings about more than 100% greater value of compressive strength (Fig. 4.25).

More anisotropic structures of polyurethane foams usually causes worsening of their heat-insulating properties when the heat flow is realized in the parallel direction to the cell elongation. In the investigation, higher thermal conductivity coefficient was estimated for the foam prepared under microwave irradiation than for the conventional foam, although the first one was characterized by more isotropic structure. This effect was caused by the lower content of closed cells in the case of polyurethane foams prepared under microwave irradiation. The cell structures of both foams prepared under microwave as well as conventional conditions are shown in Figure 4.26.

The properties of the rigid polyurethane foams blown with various physical agents were also compared. The foaming processes were carried out under microwave irradiation for 6 minutes with 75 W of microwave power. In the formulations of polyurethane foams prepared under microwave irradiation, only half the amount of the catalyst was used in relation to polyurethane systems blown conventionally. The weight amounts of different blowing agents were properly chosen in order to obtain the foams with

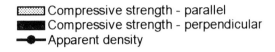

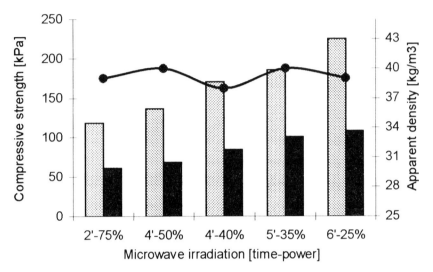

Fig. 4.25 Apparent density and compressive strength of polyurethane foams measured in parallel and perpendicular rise directions.

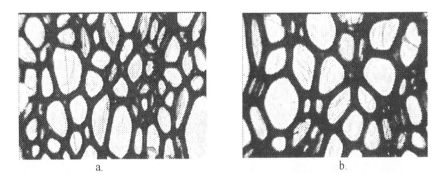

Fig. 4.26 Cell structures of rigid polyurethane foams obtained under conventional (**a**) and microwave (**b**) conditions.

similar apparent density. The mechanical properties of the foams prepared with reduced amount of the catalyst under microwave irradiation are compared in Figure 4.27.

The significant influence of foaming conditions under microwave irradiation on cell structure and properties of rigid polyurethane foams blown

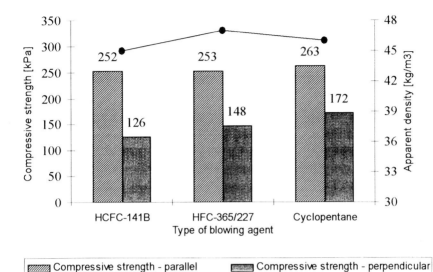

Fig. 4.27 Apparent density and compressive strength measured in parallel and perpendicular rise directions for polyurethane foams blown with different blowing agents.

with new-generation physical agents was observed. Power decrease and time elongation of microwave irradiation during foaming processes resulted in more uniform structure and better mechanical properties of rigid polyurethane foams. Eventually, the application of microwave irradiation for blowing of polyurethane foams allowed a reduction in the amount of amine catalysts. The possibilities for the application of partially decomposed cellular polyurethane waste under microwave irradiation to formulate modern ecological systems to manufacture heat-insulating foams are discussed in Chapter 5 (Bogdal et al., 2004b).

Scorching of water-blown polyurethane foams is a general phenomenon usually attributed to the thermal degradation and oxidation of the foam. Because scorch is more visible in the presence of some phosphorus- or halogen-containing fire retardants, four phosphorous-containing fire-retardant foams and one non–fire-retardant foam were tested (Table 4.8), while microwave irradiation was applied to simulate industrial foaming conditions in the lab scale (Luda et al., 2004).

Foam slabs of 30 × 30 × 20 cm were prepared, and due to better heat dissipation in the small-scale experiments, scorching usually does not appear in the lab scale. Thus, in order to provide additional heat that might better stimulate industrial foaming conditions, the samples were subjected

Table 4.8.   List of investigated polyurethane foams.

Foam symbol	Fire retardant	Chemical structure of fire retardant
NFR	None	
FR1	Fyrol PBR	Blend of pentabromodiphenyl oxide and aromatic phosphate
FR2	Fyrol FR-2	Tris(dichloroisopropyl phosphate)
FR3	AC003	Proprietary, does not contain halogen
FR4	Fyrol PNX	Oligomeric ethyl phosphate

Reprinted from Luda, M.P., Bracco, P., Costa, L., Levchik, S.V. 2004. Polym. Degrad. Stab. 83:215, with permission.

to microwave irradiation for 85 seconds in 1,000-W microwave cavity. When subjected to the microwave irradiation, a yellow-brown discoloration appears only in the FR2 and FR4 foams. Scorch looks like concentric sections of progressively lighter color from the center of the slab to the edges, which remain white. NFR and FR1 foams did not give different colors; however, when all these foams were subjected to more severe microwave treatment (i.e., 120 seconds), they all turned brown. Similar performance was observed with FR3 and FR4 foams as shown in their UV reflectance spectra of virgin and microwave-treated samples (Fig. 4.28).

Polyimides

Microwave-assisted condensation of polyimides was one of the most frequent applications in polymer chemistry. The reported work concerning preparation of polyimides under microwave conditions can be divided into three main areas:

- Polycondensation of salt monomers composed of diamines and pyromellitic acid
- Dehydration of poly(amic-acid)s as polyimide precursor
- Polymerization of nadic–end capped or phenyl ethynyl–terminated imide oligomers
- Polycondensation of imide diacid chlorides with aliphatic and aromatic amines [poly(amide imide)]

In the case of polycondensation of salt monomers, polyimides were obtained from salt monomers composed of aliphatic diamines and pyromellitic acid or its diethyl ester in the presence of a small amount of a polar organic medium (Fig. 4.29) (Imai et al., 1996a). The reactions were carried

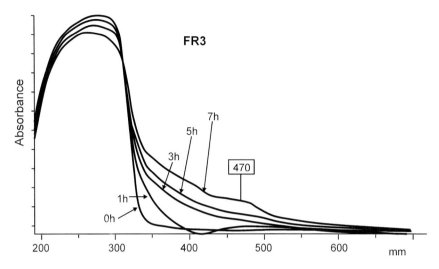

Fig. 4.28 UV spectra of FR4 foam before (white) and after (yellow and brown) microwave treatment. Reprinted from Luda, M.P., Bracco, P., Costa, L., Levchik, S.V. 2004. Polym. Degrad. Stab. 83:215, with permission.

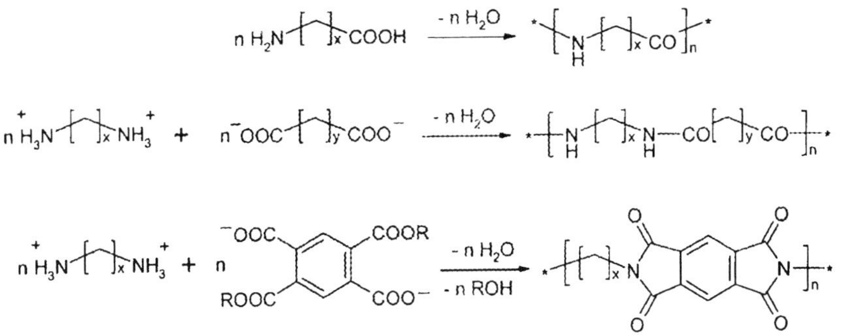

Fig. 4.29 Preparation of polyimides from salt monomers composed of aliphatic diamines and pyromellitic acid or its diethyl ester.

out in a modified domestic microwave oven with a small hole on the top of the oven so that nitrogen was introduced to 30-mL wide-mouth vial adapted as a reaction vessel.

During the experiments, a monomer salt (2 g) in a polar high boiling solvent (1 to 2 mL) that acted as a microwave absorber was irradiated under nitrogen atmosphere. The microwave-assisted polycondensation proceeded rapidly and was completed within 2 minutes for the polyimides with inher-

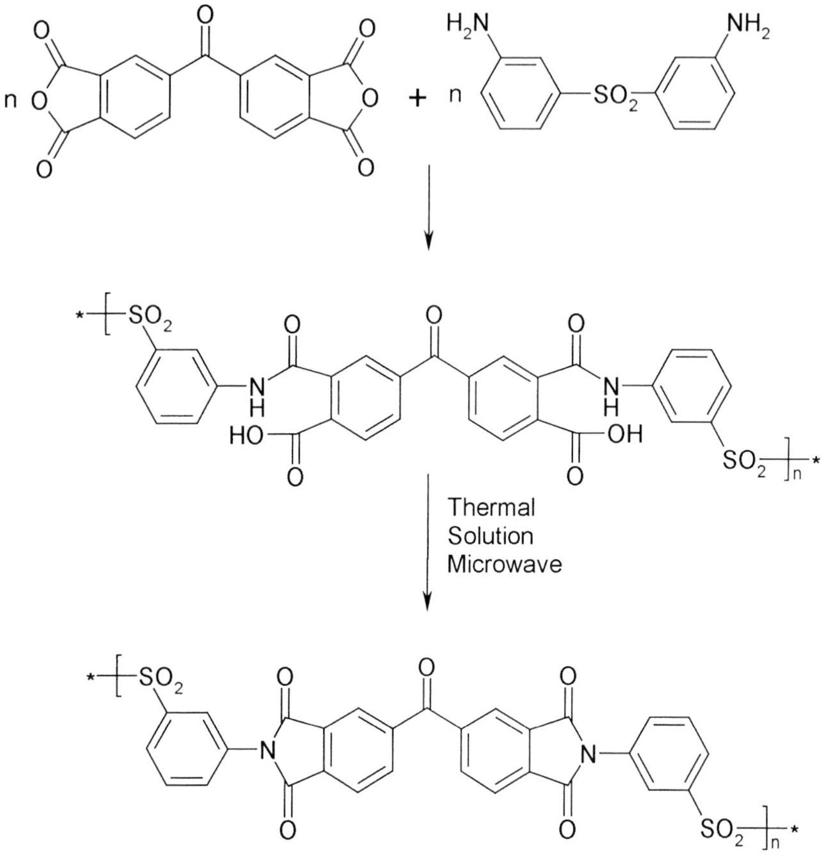

Fig. 4.30   Imidization of polyamic acid prepared from 3,3',4,4'-benzophenonte-tracarboxylic acid dianhydride (BTDA) and 3,3'-diaminodiphenyl sulfone (DDS).

ent viscosity above 0.5 dL/g (Imai et al., 1996b). Eventually, the rate of polycondensation of the salt monomers under various conditions was found to decrease in the following order: the microwave-induced polycondensation > solid-state thermal polymerization > high-pressure thermal polycondensation (Imai, 1999).

Regarding the dehydration reactions of poly(amic-acid)s as polyimide precursor, the kinetic study of imidization under microwave irradiation of polyamic acid prepared from 3,3',4,4'-benzophenontetracarboxylic acid dianhydride (BTDA) and DDS (Fig. 4.30) were carried out in a 20 wt% solution of N-methylpyrrolidone (NMP) and cyclohexylpyrrolidone (CHP), which acted as an azeotroping agent to remove the water of imidization and to prevent scission during imidization (Lewis et al., 1992).

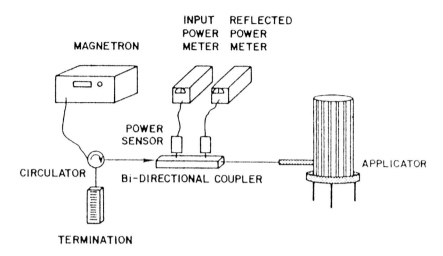

Fig. 4.31 Schematic diagram of the microwave processing equipment. Reprinted from Lewis, D.A., Summers, J.D., Ward, T.C., McGrath, J.E. 1992. J. Polym. Sci. Part A: Polym. Chem. 30:1647, with permission.

The samples were maintained in a Teflon vessel with a 1.5-cm-diameter hole and 1.5 cm depth. The microwave equipment consisted of a microwave generator (85 W) and a tunable cavity operating in the TE_{111} mode, while temperature was monitored using a fiberoptic temperature sensor (Fig. 4.31).

During a typical run, 25 W of microwave power was required to heat the sample to the desired temperature over 80 to 200 seconds. Small aliquots of the reaction mixture were taken that were sufficient for the Fourier transform-infrared analysis. Eventually, it was found that microwave irradiation acted to enhance the kinetics of solution imidization over that obtained for conventional treatment by 20 to 34 times, depending on the reaction temperature. The apparent activation energy for this imidization, determined from an Arrhenius analysis, was reduced from 105 to 55 kJ/mol when microwave activation was used rather than conventional thermal processing (Table 4.9).

Later, under microwave irradiation, polyamic acid imidization process was studied in a solid state (Chen et al., 1996). The prepolymer, polyether-ester polyamic acid, was prepared by the polycondensation of poly(tetramethylene oxide)glycol di-*p*-aminobenzoate (Polymine-650) with average molecular weight of 808 g/mol and pyromellitic acid dianhydride (PMDA) at room temperature in a DMF solution. Later, the prepolymer solution was cast on polytetrafluoroethylene plates to form 200-μm thin films that were imidized under microwave irradiation in a household microwave oven at 60 °C. The temperature was measured by means of a thermocouple applied to the film

Table 4.9. Rate constant for solution imidization.

Temperature (°C)	$k_{thermal}$ (min^{-1})	$k_{microwave}$ (min^{-1})
130		0.030
140	0.0014	
149	0.0022	
150		0.076
160		0.103
161	0.0055	
170	0.011	
175		0.169

Reprinted from Lewis, D.A., Summers, J.D., Ward, T.C., McGrath, J.E. 1992. J. Polym. Sci. Part A: Polym. Chem. 30:1647, with permission.

surface immediately after the intervals of microwave turn off. It was found that microwave irradiation reduced both the reaction temperature and time. For example, during the solid phase thermal polymerization, 68.3% polyamic acid was converted to polyimide at 155 °C, while under microwave irradiation, 65% of polyamic acid was reacted at 60 °C within 3 hours (Chen et al., 1996). In our opinion, such experiments are less reliable if temperature is only monitored in thin films during microwave turn-off cycles.

The polycondensation reactions of pyromellitic dianhydride with either benzoguanamine or 3,3'-diaminobenzophenone under microwave irradiation resulted in polyimides possessing third-order NLO properties (Fig. 4.32)

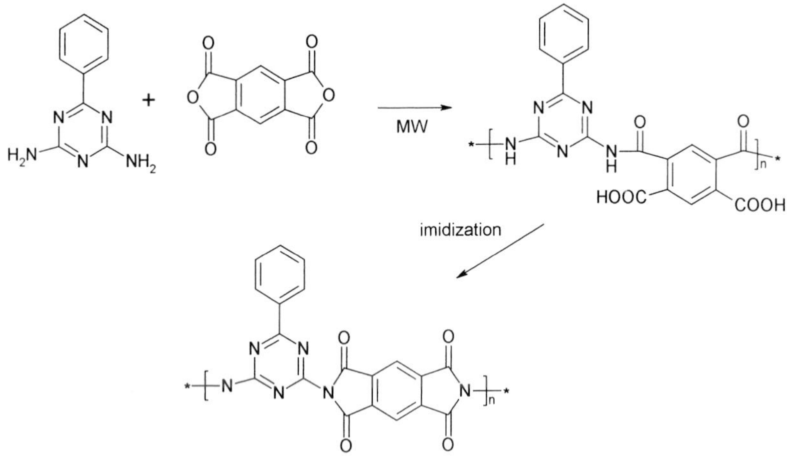

Fig. 4.32 Polycondensation reactions of pyromellitic dianhydride with benzoguanamine or 3,3'-diaminobenzophenone.

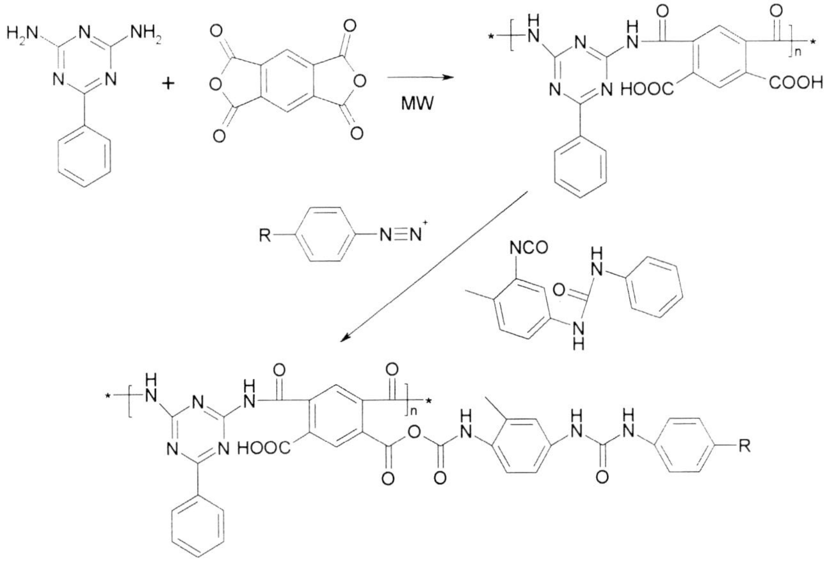

Fig. 4.33 Side chain polymers of poly(amic acid) obtained by polycondensation of benzoguanamine and pyromellitic dianhydride.

(Lu et al., 2003). The polymers obtained under microwave conditions were characterized by a large third-order nonlinearities and time response.

Side chain polymers of poly(amic acid) were obtained by polycondensation of benzoguanamine and pyromellitic dianhydride under microwave irradiation (Li et al., 2004, 2005; Lu et al., 2004). The reaction was carried out in a household microwave oven in which 100 mL of DMF solution, 33 mmol of benzoguanamine, and an equimolar amount of pyromellitic dianhydride were stirred together and irradiated for 1 hour at 60 °C (Fig. 4.33). The resulted poly(amic acid) was precipitated from the solution and then modified in order to obtained side chain polymers with fluorescent as well as third-order NLO properties.

Subsequently, the studies of the crosslinking reaction of nadic end-capped imide [i.e., *N,N'*-(oxydi-3,4'-phenylene)bis(5-norbornene-2,3-dicarboximide)] under both conventional thermal and microwave conditions were performed (Fig. 4.34) (Liu at al., 1998). The investigated starting resin (RP-46) consisted of polyimide precursors, 3,3',4,4'-benzophenontetracarboxylic acid methyl ester (BTDE), 3,4'-oxydianiline (3,4'-ODA), and end capping reagent, 5-norbornene-2,3-dicarboxylic acid monomethyl ester (NE). The cure process proceeds in two steps: (1) imidization and (2) a thermal induced (reverse Diels-Alder) decomposition-recombination cross-linking step (Fig. 4.34).

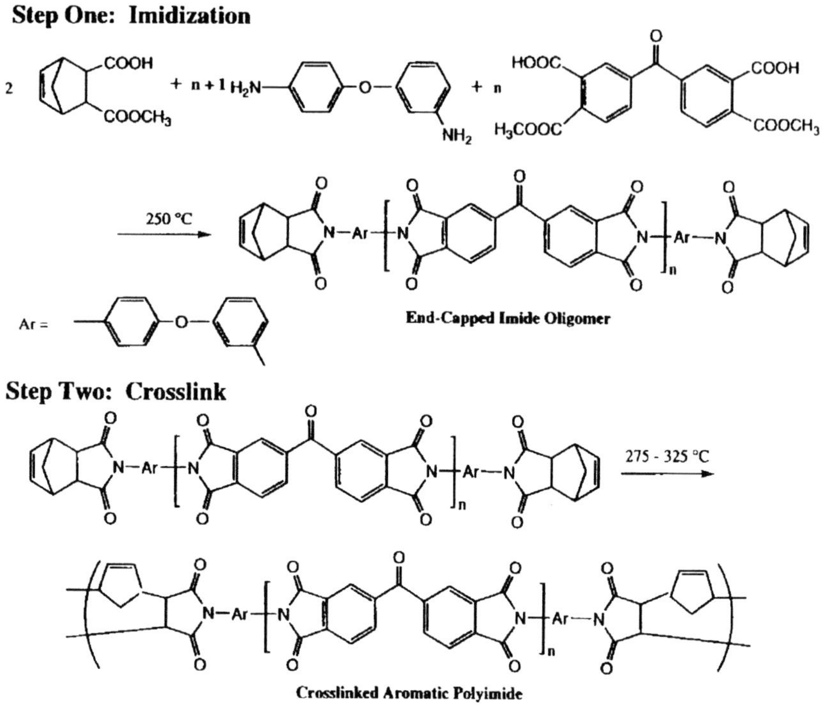

Step One: Imidization

Step Two: Crosslink

Crosslinked Aromatic Polyimide

Fig. 4.34   Crosslinking reaction of nadic end–capped imides.

Within the same time, the kinetic studies of a model compound, "bis-nadime" (Fig. 4.35), in thermal and microwave processes were carried out in order to simulate the crosslinking reaction of polyimide RP-46.

The reactions were run in a multimode microwave reactor, while the temperature was controlled by means of a thermocouple covered with a ceramic tube to avoid the influence of microwaves. The microwave cure was carried out at the temperature range of 230 to 325 °C, and the runs at 230 to 280 °C were used to determine the kinetics of the reaction. At these temperatures, the microwave cure was rapid with both isomers that underwent conversion to the crosslinked structure about 10 times faster than during a conventional thermal process (Table 4.10). The apparent activation energy for the thermal cure was estimated to be 94 kJ/mol, whereas for the microwave cure the value of activation energy fell in the range of 74 to 84 kJ/mol, which suggests that the microwave process is a more efficient energy process (Fig. 4.36) (Liu et al., 1998).

In turn, the kinetic studies of the microwave cure of phenylethynyl-terminated imide oligomer (PETI-5, M_n = 5,000 g/mol) and a model com-

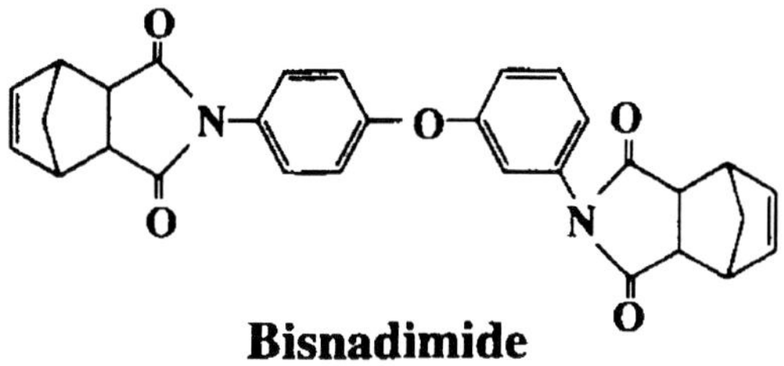

Bisnadimide

Fig. 4.35 Structure of the model compound bisnadimide.

Table 4.10. Rate constants for the thermal and microwave cure processes.

Temperature (°C)	Thermal process k (min^{-1})	Microwave process k (min^{-1})
230	0.003	0.028
280	0.011	0.140
300	0.030	
325	0.128	
Activation energy (kJ/mol)	94	84

Reprinted from Liu, Y., Sun, X.D., Xie, X.Q., Scola, D.A. 1998. J. Polym. Sci. Part A: Polym. Chem. 36:2653, with permission.

pound, 3,4′-bis[(4-phenylethynyl)phthalimido] diphenyl ether (PEPA-3,4′-ODA) (Fig. 4.37), were also performed (Fang et al., 2000a).

The microwave cure of PEPA-3,4′-ODA and PETI-5 oligomer was performed in a variable frequency (4.19 to 5.19 GHz) microwave oven in the temperature range of 300 to 330 °C and 350 to 380 °C, respectively. The powder sample was placed in a glass tube, and the tube was immersed in silicon carbide powder, as shown in Figure 4.38.

Compared with thermal kinetic curing rate studies of PEPA-3,4′-ODA and PETI-5, microwave cure gave much higher rate constants for both. For PEPA-3,4′-ODA, the reaction followed first-order kinetics, yielding an activation energy of 27.6 kcal/mol, which was 68% that of the thermal cure. For PETI-5, the reaction followed 1.5 order, yielding an activation energy of 17.1 kcal/mol, which was 51% that of the thermal cure for PETI-5 (Table 4.11).

Eventually, the kinetic data showed that maintaining the same temperature with microwave irradiation could provide a faster cure relative to con-

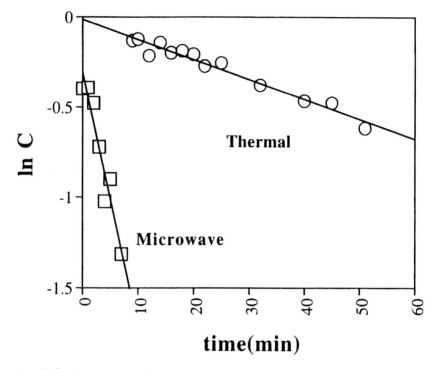

Fig. 4.36 Comparison of the reaction rate for microwave and thermal cure of the bisnadimide model compounds at 280 °C. Reprinted from Liu, Y., Sun, X.D., Xie, X.Q., Scola, D.A. 1998. J. Polym. Sci. Part A: Polym. Chem. 36:2653, with permisssion.

ventional thermal heating or the same cure rate provided that the curing temperature under microwave irradiation is adequately lower (Table 4.12) (Fang et al., 2000b).

The microwave-assisted polycondensation of aromatic diisocyanates (i.e., *p*-phenylene disocyanate [PPDI] and 1,5-dinaphthalene diisocyanate [NDI]) and dianhydrides (i.e., pyromellitic dianhydride [PMDA] and 3,3′,4,4′-benzophenontetracarboxylic acid dianhydride [BTDA]) for preparation of polyimide was also studied (Fig. 4.39) (Yeganeh et al., 2004).

The reaction was carried out on a Teflon reactor in which dianhydride (1 mmol), diisocyanate (1 mmol), and solvent mixed with a catalyst were irradiated in a domestic microwave oven without a temperature control. Immediately after the reaction, the polymer formed was isolated by pouring into methanol (100 mL) and purified by refluxing it for 2 hours.

It has been found that a polar organic medium was necessary to induce

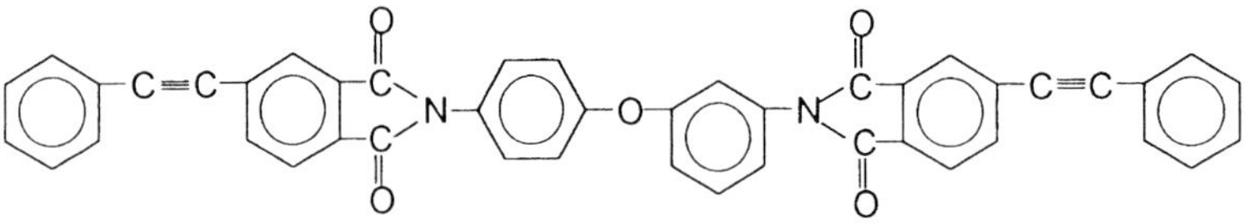

**3,4'-Bis[4-(phenylethynyl)phthalimido]diphenyl Ether
(PEPA-3,4'-ODA)**

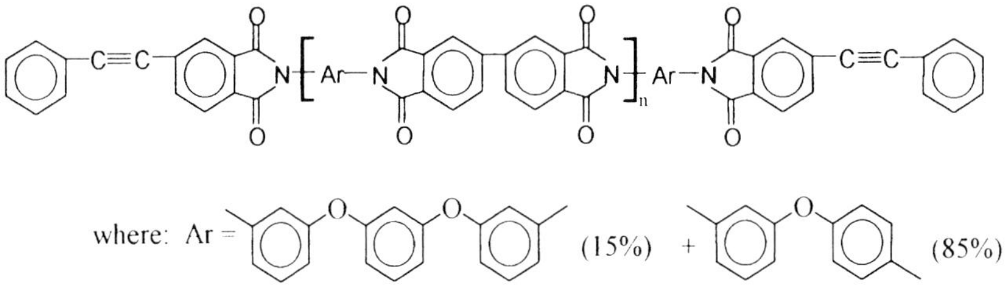

where: Ar = (15%) + (85%)

**Phenylethynyl-terminated Imide Oligomer (MW ~ 5,000 g/mol)
(PETI-5)**

Fig. 4.37 Chemical structures of PEPA-3,4'-ODA and PETI-5.

an effective homogeneous reaction mixture and, hence, cause efficient polycondensation to achieve polyimides. The data concerning the dependence of polymer inherent viscosity to the nature of solvent are presented in Table 4.13. It was found that reactions performed in NMP produced the polymer with the highest inherent viscosity and molecular weights of polymers were increased with increase of boiling points of the solvents.

Finally, it was concluded that microwave-assisted polymerization of dianhydrides and diisocyanates is a rapid and facile method for the preparation of polyimides. Using a closed Teflon mold, containing 1 mmol of each reactant, NMP solvent, 0.1 mmol of tributylamine catalyst, 70% solid content of monomers, and 15-minute microwave irradiation with full (100%) power, polymers with the highest inherent viscosity were obtained (Yeganeh et al., 2004).

Poly(amide imide)s were obtained by the polycondensation of a number of diacid chlorides such as N,N'-(4,4'-carbonyldiphthaloyl)-bisalanine diacid chloride (Mallakpour et al., 2000a, 2001a, 2001b) and 4,4'-(hexafluoroiso-propylidene)-N,N'-bis(phthaloyl-L-leucine)-diacid chloride (Mallakpour et al., 2000b, 2000c) with certain aromatic amines (Fig. 4.40).

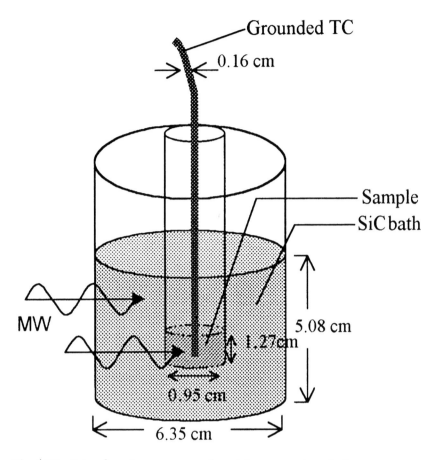

Fig. 4.38   Setup for microwave cure of model compounds and oligomer. Reprinted from Fang, X., Hutcheon, R., Scola, D.A. 2000a. J. Polym. Sci. Part A: Polym. Chem. 38:2526, with permission.

The reactions were performed in a domestic microwave oven that was used without any modification. Prior to the microwave irradiation, 0.1 g of diacid chloride was ground with an equimolar amount of an aromatic amine or diphenol and a small amount of a polar high boiling solvent (e.g., *o*-cresol, 0.05 to 0.45 mL) that acted as a primary microwave absorber. Under microwave irradiation, the polycondensation reactions proceeded rapidly (6 to 12 minutes) compared with a conventional solution polymerization (reflux for 12 hours in chloroform and then for another 12 hours in dimethylacetamide solutions (Mallakpour et al., 2000c) to give polymers with higher inherent viscosities in the range of 0.36 to 1.93 dL/g (Table 4.14).

Table 4.11. Rate constants for the thermal and microwave cure of PEPA-3,4'-ODA and PETI-5.

	Temperature (°C)	Rate constant (min^{-1})	Regression coefficient
PEPA-3,4'-ODA			
Microwave cure	300	0.02504	0.932
	310	0.03854	0.992
	320	0.06313	0.990
	330	0.08079	0.957
Thermal cure	318	0.00157	0.991
	336	0.00341	0.994
	355	0.01229	0.997
	373	0.02744	0.983
PETI-5			
Microwave cure	350	0.161	0.967
	360	0.198	0.966
	370	0.241	0.937
	380	0.306	0.915
Thermal cure	350	0.0745	0.911
	360	0.110	0.978
	370	0.152	0.988
	380	0.273	0.970
	390	0.371	0.953

Reprinted from Fang, X., Hutcheon, R., Scola, D.A. 2000a. J. Polym. Sci. Part A Polym. Chem. 38:2526, with permission.

Table 4.12. Comparison of activation energies of microwave and thermal cure reactions.

Sample	MW	Δ
PEPA-3,4'-ODA	27.6 ± 2.3 (kcal/mol)	40.7 ± 2.7 (kcal/mol)
PETI-5	1.20 ± 0.10 (eV)	1.77 ± 0.12 (eV)
	17.1 ± 0.7 (kcal/mol)	33.8 ± 2.0 (kcal/mol)
	0.74 ± 0.03 (eV)	1.47 ± 0.09 (eV)

Reprinted from Fang, X., Hutcheon, R., Scola, D.A. 2000a. J. Polym. Sci. Part A Polym. Chem. 38:2526, with permission.

In the case of the polycondensation of 4,4'-(hexafluoroisopropylidene)-*N,N'* bis(phthaloyl L leucine) diacid chloride with aromatic amines, the thermal stability of the resulted polymers showed no significant difference from the polymers obtained in conventional solution polymerization, but

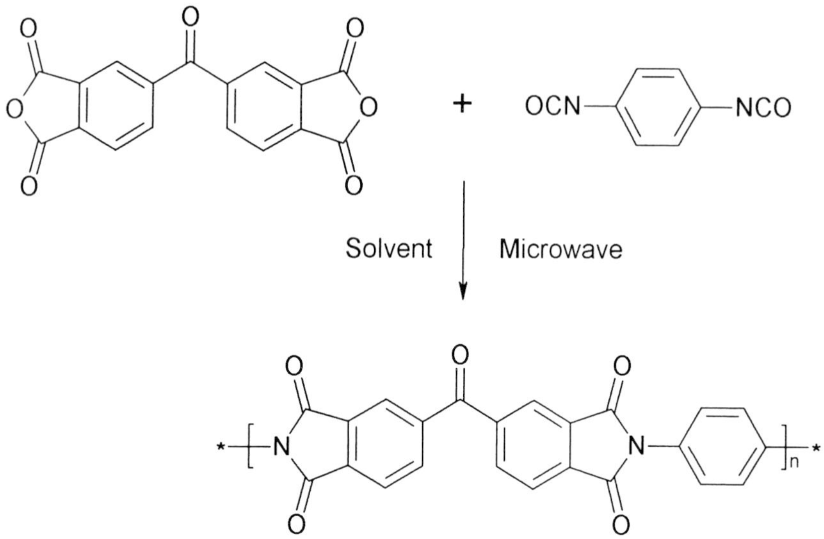

Fig. 4.39 Polycondensation of aromatic diisocyanates and dianhydrides.

Table 4.13. Dependence of polyimides inherent viscosity to solvent type, with 1 mmol of each BTDA and PPDI, 0.25 g of solvent, and 15-minute irradiation at full power.

Solvent	ηinh, dL/g	Yield, %
DMF (b.p. 153 °C)	0.163	91
DMAc (b.p. 166 °C)	0.200	96
DMSO (b.p. 189 °C)	0.218	95
NMP (b.p. 202 °C)	0.308	98

Reprinted from Yeganeh, H., Tamami, B., Ghazi, I. 2004. Eur. Polym. J. 40:2059, with permission.

higher glass-transition temperatures (approximately 19 to 82 °C) were reported (Mallakpour et al., 2000c). Similar series of poly(amide imide)s were obtained by the polycondesation of diacid chlorides such as *N,N*-(4,4′-carbonyldiphthaloyl)-bis-isoleucine diacid chloride (Mallakpour et al., 2004a), *N,N*′-(4,4′-carbonyldiphthaloyl)-bis-l-phenylalanine)-diacid chloride (Mallakpour et al., 2002), and *N,N*′-(4,4′-sulfonediphthaloyl)-bis-l-phenylalanine)-diacid chloride (Mallakpour et al., 2003a) with certain aromatic amines (Fig. 4.41), and diacid chlorides derived from Epiclon B-4400

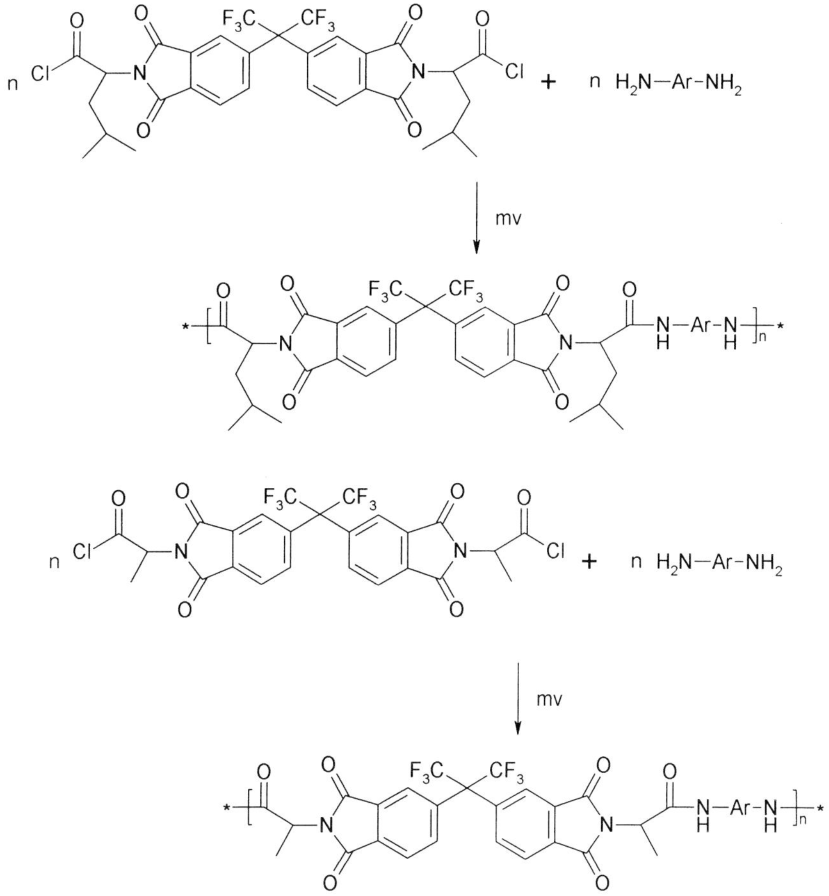

Fig. 4.40 Polycondensation of diacid chlorides with certain aromatic amines.

and phenylalanine (Mallakpour et al., 2004b) or L-luecine (Mallakpour et al., 2003b) with aromatic amines as well (Fig. 4.42).

Under microwave irradiation, the polycondensation reactions proceeded rapidly (6 to 12 minutes) compared with a conventional solution polymerization (i.e., reflux for 5 to 10 hours in chloroform or NMP) to give a series of optically active polymers with inherent viscosities in the range of 0.12 to 0.52 dL/g.

Another series of poly(amide imide)s were obtained by polycondensation of hydantoin and thiohydantoin derivatives of pyromellitic acid chlorides with N,N'-(4,4'-carbonyldiphthaloyl)-bisalanine diacid chloride

Table 4.14.   Inherent viscosity (γinh) of the polymer (differences between the solution and microwave methods.

Reagent	Solution η_{inh}	Microwave η_{inh}
Benzidine	0.22	1.22
4,4′-Diaminodiphenylmethane	0.29	1.32
1,5-Diaminoantraquinone	0.09	1.73
4,4′-Sulfonyldianiline	0.09	0.50
3,3′-Diaminobenzophenone	0.22	1.26
p-Phenylenediamine	0.15	1.04
2,6-Diaminopyridine	0.12	1.44

[a] One equimolar of diacid chloride and diamines was refluxed for 12 h in CHCl3 and then heated for 12 h at 120 °C in dimethylacetamide.

Reprinted from Mallakpour, S.E., Hajipour, A.R., Khoee, S. 2000c. J. Polym. Sci. Part A Polym. Chem. 38:1154, with permission.

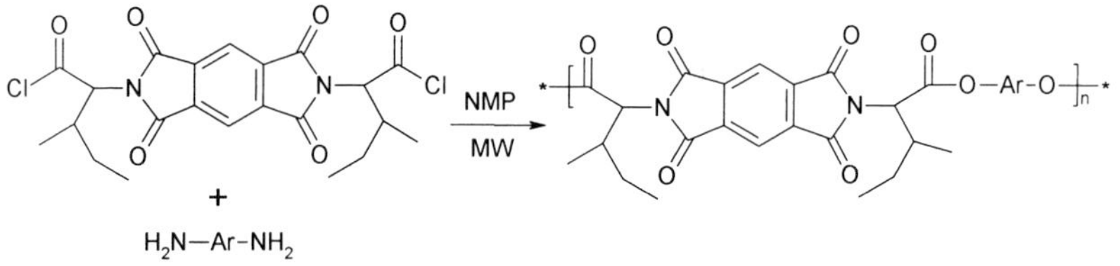

Fig. 4.41   Polycondesation of diacid chlorides with some aromatic amines.

(Faghihi et al., 2003), *N,N′*-(pyromellitoyl)-bis-l-phenylalanine diacid chloride (Faghihi et al., 2004a), and *N,N′*-(4,4′-diphenylether)-bistrimellitide diacid chloride (Faghihi et al., 2004b) prepared under microwave irradiation (Fig. 4.43). Typically, the polymerization reactions were run in a domestic microwave oven for 10 minutes without temperature control, while 1.0 mmol of a diacid chloride was mixed with an equimolar amount of hydantoin derivatives in the presence of 1 mL of *o*-cresol. The resulting poly(amide imide)s were obtained in good yields with inherent viscosities of approximately 0.28 to 0.66 dL/g.

In all the cases of reactions performed in a household microwave oven, it is difficult to make a direct comparison of reaction rates for conventional and microwave methods because the reactions were run under different reaction conditions (i.e., solvent-less reaction and reaction in a solution for microwave and conventional conditions, respectively). Moreover, the temperatures for all the microwave experiments were not measured.

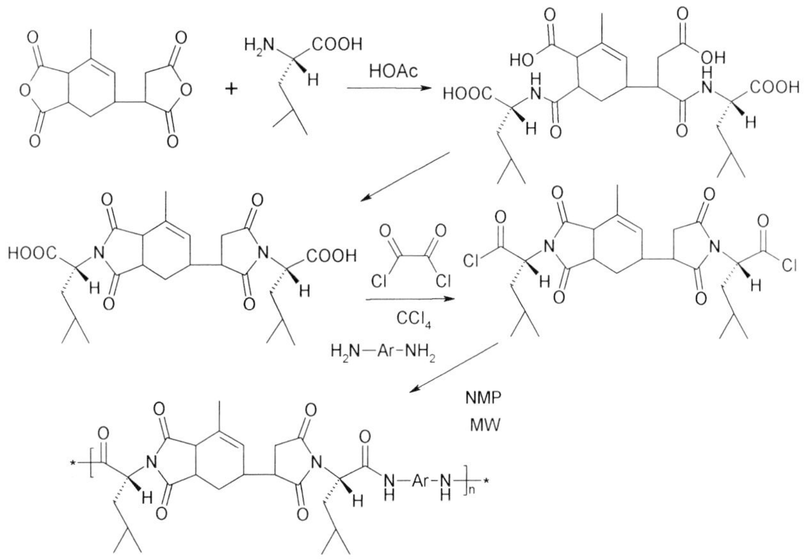

Fig. 4.42   Polycondesation of diacid chlorides with some aromatic amines.

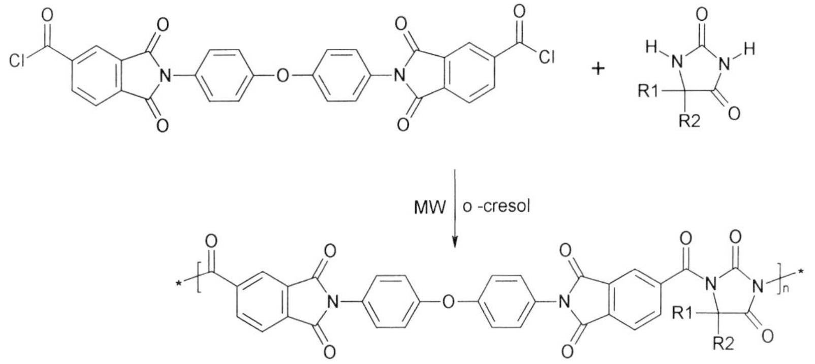

Fig. 4.43   Polycondesation of hydantois and thiohydantoins derivatives of pyromellitic acid chlorides.

Likely, poly(amide imide) poly(ester imide)s were obtained by the polycondensation of a number of diacid chlorides such as *N,N'*-(4,4'-carbonyldiphthaloyl)-bisalanine diacid chloride (Mallakpour et al., 2000a), 4,4'-(hexafluoroisopropylidene)-*N,N'*-bis(phthaloyl-L-leucine)-diacid chloride (Mallakpour et al., 2000b), as well as *N,N'*-(pyrromellitoyl)-bis-

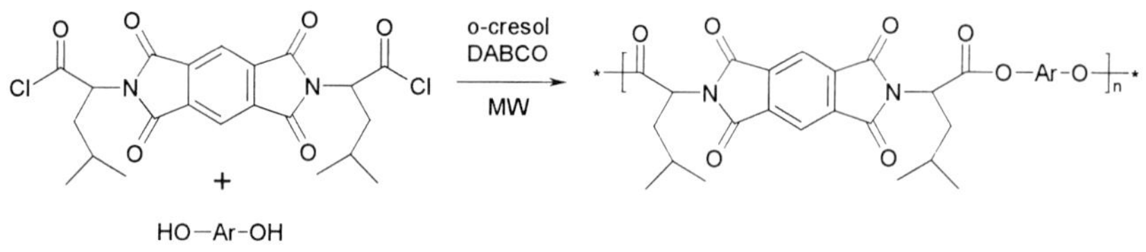

Fig. 4.44 Polycondensation of diacid chlorides with certain and biphenols.

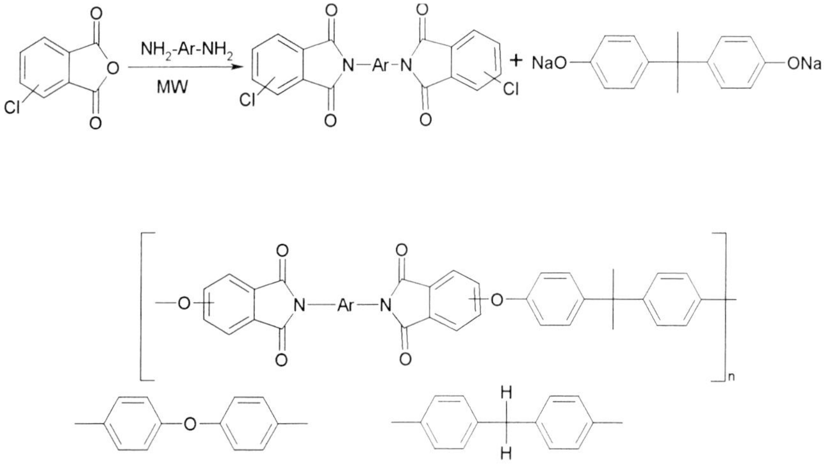

Fig. 4.45 Polycondensation of disodium salt of bisphenol-A and bis(chlorophthalimide)s.

L-leucine diacid chloride (Mallakpour et al., 2000c) with certain bisphenols (Fig. 4.44).

For instance, in the last case, the reaction route involved the reaction of pyromellitic anhydride with L-leucine, then conversion of the resulted diacid into the diacid chloride, which in turn reacted with a number of diols such as phenol phthalein, bisphenol A, 4,4′-hydroquinone, etc., under microwave irradiation (Fig. 4.44). The polymerization reactions were carried out within 10 minutes in a domestic microwave oven in a porcelain dish in which 0.10 g of diacid chloride was mixed with an equimolar amount of diol in the presence of small amounts of o-creasol and DABCO as catalysts. As a result, a series of optically active poly(ester imide)s were obtained in a good yield and moderate inherent viscosity of 0.10 to 0.27 dL/g.

The synthesis of poly(ether imide)s in the condensation of disodium salt of bisphenol A and bis(chlorophthalimide)s under microwave irradiation was also described (Fig. 4.45) (Gao et al., 2004). The polymerization reactions were performed under phase-transfer catalysis (PTC) conditions in an *o*-dichlorobenze solution. For this purpose, 16.12 mmol of bis(chloro-phthalimide)s and 16.12 mmol of disodium salt of bisphenol A in 60 mL of *o*-dichlorobenze with 0.56 mmol of hexaethyl guanidinium bromide was ir-radiated in a domestic microwave oven for 25 minutes and then precipi-tated from methanol. The polymerization reactions, in comparison with conventional heating polycondensation, proceeded rapidly (25 minutes versus 4 hours at 200 °C), and the polymers with inherent viscosities in the range of 0.55 to 0.90 dL/g were obtained.

References

Beldjoudi, N., Gourdenne, A. 1988. Eur. Polym. J. 24:265.

Boey, F.Y.C., Rath, S.K. 2000. Adv. Polym. Techn. 19:194.

Boey, F.Y.C., Yap, B.H. 2001. Polym. Test.20:837.

Boey, F.Y.C., Yap, B.H., Chia, L. 1999. Polym. Test. 18:93.

Bogdal, D., Gorczyk, J. 2003. Polymer. 44: 7795.

Bogdal, D., Gorczyk, J. 2004. J. Appl. Polym. Sci. 94:1969.

Bogdal, D., Gorczyk, J., Brzozowski, Z.K., Staszczak, S.K., Hadam, L.K., Rupinski, S. 2006. J. Appl. Polym. Sci. 100:3850.

Bogdal, D., Prociak, A., Pielichowski, J. 2004b. Application of Microwave Irradiation for Chemical Recycling of Polyurethanes. In: Proceedings of Global Symposium on Recycling, Waste Treatment and Clean Technology, REWAS 2004, Madrid, vol. 3, pp 2807-2809.

Casalini, R., Corezzi, S., Livi, A., Levita, G., Rolla, P.A. 1997. J. Appl. Polym. Sci. 65:17.

Chen, J., Chen, Q., Yu, X. 1996. J. Appl. Polym. Sci. 62:2135.

Chen, M., Siochi, E.J., Ward, T.C., McGrath, J.E. 1993. Polym. Eng. Sci. 33:1092.

Delmotte, M., Jullien, H., Ollivon, M. 1991. Eur. Polym. J. 27:371.

Faghihi, K., Hajibeygi, M. 2004b. J. Appl. Polym. Sci. 92:3447.

Faghihi, K., Zamani, K., Mirsaamie, A., Mallakpour, S. 2004a. J. Appl. Polym. Sci. 91:516.

Faghihi, K., Zamani, K., Mirsaamie, A., Sangi, M.R. 2003. Eur. Polym. J. 39:247.

Fang, X., Hutcheon, R., Scola, D.A. 2000a. J. Polym. Sci. Part A: Polym. Chem. 38:2526.

Fang, X., Hutcheon, R., Scola, D.A. 2000b. J. Polym. Sci. Part A: Polym. Chem. 38:1379.

Gao, C., Zhang, S., Gao, L., Ding, M. 2004. J. Appl. Polym. Sci. 92:2414.

Hedreul, C., Galy, J., Dupuy, J., Delmotte, M., More, C. 1998. J. Appl. Polym. Sci. 68:543.

Imai, Y. 1996a. A new facile and rapid synthesis of polyamides and polyimides by microwave assisted polycondensation. In: Step-Growth Polymers for High Perfor-mance Materials: New Synthetic Methods, edited by James L. Hendrick and Jeff W. Labadie, pp. 421:430, ACS, Series No 624.

Imai, Y. 1999. Adv. Polym. Sci. 140:1

Imai, Y., Nemoto, H., Kakimoto, M. 1996b. J. Polym. Sci. Part A: Polym. Chem. 34:701.

Jordan, C., Galy, J., Pascault, J.P., More, C., Delmotte, M., Jullien, H. 1995. Polym. Eng. Sci. 35:233.

Jow, J., DeLong, J.D., Hawley, M.C. 1989. SAMPE Quart 20:46.

Jullien, H., Valot, H. 1985. Polymer 26:506.

Levita, G., Livi, A., Rolla, P.A., Culicchi, C. 1996. J. Polym. Sci. Part B: Polym. Phys. 34:2731.

Lewis, D.A., Summers, J.D., Ward, T.C., McGrath, J.E. 1992. J. Polym. Sci. Part A: Polym. Chem. 30:1647.

Li, N., Lu, J., Yao, S. 2005. Macromol. Chem. Phys. 206:559.

Li, N., Lu, J., Yao, S. Xia, X., Zhu, X. 2004. Materials Lett. 58:3115.

Liu, Y., Sun, X.D., Xie, X.Q., Scola, D.A. 1998. J. Polym. Sci. Part A: Polym. Chem. 36:2653.

Lu, J., Yao, S., Tang, X., Sun, M., Zhu, X. 2004. Opt. Mat. 25:359.

Lu, J.M., Ji, S.J., Chen, N.Y., Zhang, Z.B., Sun, Z.R., Zhu, X.L., Shi, W.P. 2003. J. Appl. Polym. Sci. 87:1739.

Luda, M.P., Bracco, P., Costa, L., Levchik, S.V. 2004. Polym. Degrad. Stab. 83:215.

Mallakpour, S.E., Hajipour, A.R., Khoee, S. 2000c. J. Polym. Sci. Part A: Polym. Chem. 38:1154.

Mallakpour, S., Hajipour, A.R., Zamanlou, M.R. 2003b. J. Polym. Sci. Part A: Polym. Sci. 41:1077.

Mallakpour, S., Hajipour, A.R., Zamanlou, M.R., 2002. Eur. Polym. J. 38:475.

Mallakpour, S., Kowsari, E. 2003a. J. Polym. Sci. Part A: Polym. Chem. 41:3974.

Mallakpour, S., Shahmohammadi, M.H. 2004a. J. Appl. Polym. Sci. 92:951.

Mallakpour, S., Zamanlou, M.R. 2004b. J. Appl. Polym. Sci. 91:3281.

Mallakpour, S.E., Hajipour, A.R., Faghihi, K. 2000a. Polym. Int. 49:1388.

Mallakpour, S.E., Hajipour, A.R., Faghihi, K. 2001a. Eur. Polym. J. 37:119.

Mallakpour, S.E., Hajipour, A.R., Khoee, S. 2000b. J. Appl. Polym. Sci. 77:3003.

Mallakpour, S.E., Hajipour, A.R., Zamanlou, M.R. 2001b. J. Polym. Sci. Part A: Polym. Chem. 39:177.

Marand, E., Baker, H.R., Graybeal, J.D. 1992. Macromolecules 25:2243.

Mijovic, J., Corso, W.V., Nicolais, L., d'Ambrosio, G. 1998. Polym. Adv. Technol. 9:231.

Mijovic, J., Fishbain, A., Wijaya, J. 1992. Macromolecules 25:986.

Parodi, F. 1999. Microwave Heating and the Acceleration of Polymerization Processes. In: Polymers and Liquid Crystals, edited by Andrzej Wlochowicz, Proceedings of SPIE - The International Society for Optical Engineering 4017:2.

Parodi, F., Gerbelli, R., De Meuse, M. 1996. US Patent 5 489 664.

Parodi, F., Belgiovine, C., Zannoni, C. 1994. US Patent 5 288 833.

Prociak, A., Bogdal, D., Pielichowski, J., Dziadczyk, J. 2006b. Rigid polyurethane foams blown under microwave irradiation. Proceedings of the 8th International Conference "Blowing Agents and Foaming Processes 2006" Munich, Germany.

Prociak, A. 2004. Modern polyurethane based cellular materials for heat insulating applications. In: Modern Polymeric Materials for Environmental Applications, edited by Krzysztof Pielichowski, pp. 121-124, Krakow.

Prociak, A. 2006a. Rigid polyurethane foams modified with vegetable oil-based polyols. In: Proceedings of UTECH 2006 Conference, Maastricht, The Netherlands.

Randall, D., Lee, S. 2002. The Polyurethanes Book. Wiley & Sons.

Silinski, B., Kuzmycz, C., Gourdenne, A. 1987. Eur. Polym. J. 23:273.

Thuillier, F.M., Jullien, H. 1989. Makromol. Chem., Macromol. Symp. 25:63.

Verchere, D., Sautereau, H., Pascault, J.P., Moschiar, S.M., Riccardi, C.C., Williams, R.J.J. 1990. J. Appl. Polym. Sci. 41:467.

Wei, J., Hawley, M.C., DeLong, J.D., Demeuse, M. 1993. Polym. Eng. Sci. 33:1132.

Yeganeh, H., Tamami, B., Ghazi, I. 2004. Eur. Polym. J. 40:2059.

Zhang, D., Crivello, J.V., Stoffer, J.O. 2004. J. Polym. Sci. Part B: Polym. Phys. 42 : 4230.

Zhou, J., Shi, C., Mei, B., Yuan, R., Fu, Z. 2003. J. Mater. Process. Technol. 137:156.

Zong, L., Kempel L.C., Hawley, M.C. 2005. Polymer. 46:2638.

5

POLYMER COMPOSITES AND BLENDS

Composite materials are prepared in order to improve selected properties of polymers. In the preparation process, heat is needed to liquefy thermoplastic polymer and to cure monomers or prepolymers. The possibility of very rapid curing of heterogeneous materials and the formation of unique structure has been presented by many workers. Radiation processing, including microwave heating, is an economical and applicable method of physical and chemical modification of polymeric materials. Beneficial properties of composite materials have been obtained, especially for such polymers as epoxies, polyesters, polyurethanes, and their derivatives.

Composites of Selected Resins

Functionally graded materials are designed in order to reduce thermal stress and improve mechanical properties, heat resistance, and ductility. The glass-transition temperature and modulus of graded materials in an "epoxy plastic–polyurethane elastomer" system prepared under microwave were reported (Liu et al., 2004). The objective of this investigation was microwave heating application for curing epoxy (EP) and polyurethane (PUR) resins with the addition of diaminodiphenyl methane (DDM) as a crosslinking agent, of both EP and isocyanate in PUR. Different mass ratios of EP and PUR were mixed with a stoichiometric amount of DDM. The microwave curing was conducted in a cavity connected to 2.45-GHz magnetron with variable power. The thermal curing was conducted at 120 °C. The example of specimen temperature versus time during microwave and conventional curing is shown in Figure 5.1.

The mechanical properties of the system with EP/PUR ratio of 10:10 w/w carried under different conditions are shown in Table 5.1.

The properties of the sample were related only to the microwave power setting and prolonging the irradiation time did not influence the tensile

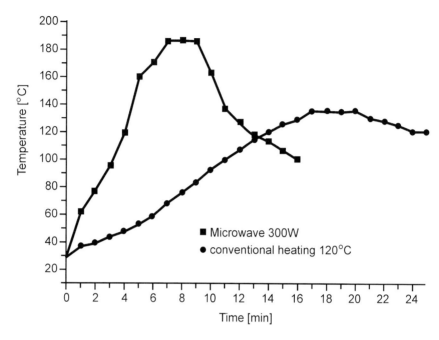

Fig. 5.1.   Specimen temperatures versus cure time during microwave and thermal curing for EP/PU/DDM system (EP/PU = 10/10 w/w). Reprinted from Liu, X.Q., Wang, Y.S., Zhu, J.H. 2004. J. Appl. Polym. Sci. 94:994, with permission.

strength, modulus, and elongation. Although the curing time of the specimen cured under microwave irradiation at 400 W was shorter, the material had better mechanical properties in comparison to the sample cured conventionally (Liu et al., 2004).

The possibility of designing and obtaining functionally graded materials based on EP/PUR by changing the content of components is presented in Table 5.2.

A solution containing 65 wt% of EP/PUR/DDM mixture in dichloromethylene was poured onto the polytetrafluoroethylene (PTFE) mold and then irradiated at 200 W of microwave power for 20 minutes. The film was submitted for the next casting. After each layer was poured in, the whole sample was successively irradiated at 400 W for 30 minutes. Applying this procedure, the materials varied gradually in the glass temperature from −54 to 162 °C, and modulus from 0.069 to 3.20 GPa was obtained (Liu et al., 2004).

Microwave heating has been examined numerically and experimentally as an alternative to conventional thermal processing techniques and investigated for a glass/epoxy laminate (Thostenson et al., 1997). A numerical

Table 5.1. Mechanical properties of the EP/PU/DDM system cured with microwave irradiation and conventional heating.

Curing cycle	E (MPa)	Tensile strength (MPa)	Tensile strain (%)
Microwave			
200 W/40 min	782	19.2	4.3
200 W/60 min	783	19.4	4.4
200 W/120 min	782	19.3	4.3
200 W/20 min + 400 W/30 min	1011	25.8	8.08
200 W/20 min + 400 W/60 min	1013	25.2	8.06
Heating:			
120 °C/360 min	972	24.9	6.53

EP/PU = 10/10 w/w.
Reprinted from Liu, Y., Xiao, Y., Sun, X., Scola, D.A. 1999. J. Appl. Polym. Sci. 73:2391, with permission.

Table 5.2. Properties along the thickness direction in the FGM.

Thickness direction in the FGM										
Layer	1	2	3	4	5	6	7	8	9	10	11
PU/EP (w/w)	10/0	10/2	10/4	10/6	10/8	10/10	8/10	6/10	4/10	2/10	0/10
Tensile strength (MPa)	4.65	5.84	11.6	15.2	20.8	25.8	27.4	32.5	45.8	75.9	64.8
T_g (°C)	−54	−9.3	25.4	45.7	83.2	99	109.2	113.2	139.6	145	162
					44.8	60	39.4				
E (GPa)	0.069	0.078	0.245	0.86	0.99	1.01	1.438	1.585	2.62	2.72	3.20

Microwave curing cycle: 200 W/20 min + 400 W/30 min.
Reprinted from Liu, Y., Xiao, Y., Sun, X., Scola, D.A. 1999. J. Appl. Polym. Sci. 73:2391, with permission.

simulation has been developed to predict the one-dimensional transient temperature profile of the composite during both microwave irradiation and conventional heating. Numerical and experimental results were presented for a glass/epoxy laminate with a thickness of 25 mm. The raw materials used for the experimental investigation were a biaphenol F/ epichlorohydrin epoxy resin (Shell Epon 862) and an aromatic diamine (Shell Epi-Cure W) as a curing agent. The microwave applicator was a cylindrical multimode cavity with an internal volume of 500 L. The large cavity was equipped with a mode stirrer and multiple microwave inputs to enhance the uniformity of the microwave field. The output power of the microwave source was varied continuously from 0 to 6 kW. The results showed that it was possible to cure thick glass/epoxy composites uniformly and eliminate temperature excursions due to chemical reactions during the cure cycle. In addition, through continuous wave feedback control of the microwave power, it was possible to monitor the cure behavior of the composite. In conclusion, microwave processing has the potential to increase the quality of thick-section composites and reduce the manufacturing costs through more efficient transfer of energy (Thostenson et al., 1997).

In the later stage, a calorimetric analysis (DSC) of the cure kinetics of the same glass/epoxy composite was conducted for both thermal and microwave curing samples (Thostenson et al., 1998). To develop a kinetic model for conventional thermal cure, isothermal experiments were conducted in the range of 135 to 175 °C, and the degree of cure as a function of time was calculated. A different approach was required to examine the cure kinetics under microwave conditions because the cure behavior was hard to monitor in situ within a microwave cavity. Therefore, samples that were placed in a microwave cavity were heated rapidly to the desired cure temperature, and the cure was stopped at a specified time by removing the sample from the cavity and quenching it. Then, the samples were analyzed by means of DSC to obtain the residual heat of reaction. It was found that the reaction rate was accelerated by the application of microwave irradiation. The results of the numerical simulation showed that it was possible to decrease the cycle times for processing of thick-section composites because microwave heating allowed for a better control of the spatial solidification of the laminate, which resulted in significantly reduced processing times and enhanced quality (Thostenson et al., 1998).

Finally, this work summarized that both numerical and experimental results indicated that volumetric heating due to microwaves promoted an inside-out cure of the thick laminates and dramatically reduced the overall processing time (Thostenson et al., 2001). Under conventional thermal conditions, to reduce thermal gradients, thick laminates were processed at lower cure temperature and heated with slow heating rates, resulting in excessive cure times. Outside-in curing of the autoclave-processed composite resulted in visible matrix cracks, while cracks could not be seen in the microwave-processed composite. The formation of cure gradients within the two composites cured under both microwave and conventional conditions is presented in the Figure 5.2.

Although cure gradients exist in both composites treated under microwave and thermal conditions, the differences in the solidification behavior are seen. In the conventionally processed composite, the outside-in cure

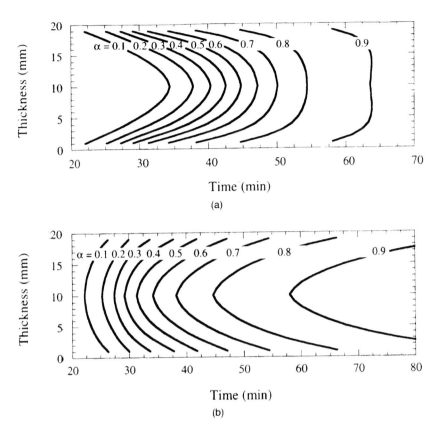

Fig. 5.2. Formation of cure gradients with two laminates during (**a**) conventional and (**b**) microwave cures. Reprinted from Thostenson, E.T., Chou, T.-W. 2001. Polym. Composites, 22:197, permission.

gradients are most significant during the early stages of the cure cycle, and the maximum cure rate for this epoxy resin system occurs at the beginning of cure. Therefore, it is critically important to initiate an inside-out cure at the beginning of the cure cycle. Reduced thermal gradients during the early stages of microwave curing allow for better control over solidification behavior of the resin. In conventional processing, very slow heating rates are required to reduce the thermal lag and heat the composite up to a temperature where additional heat is generated by the chemical reaction. Once additional heat is generated, it will help to promote the desired inside-out cure. To obtain an inside-out cure in conventional processing, the required cycle time was almost three times longer than in the case of microwave processing. Thus, the processing time can be drastically reduced to achieve the desired inside-out cure through the use of microwaves (Thostenson et al., 2001).

In another example, the fiber-glass/epoxy (Dow-Derakane 411-350) composite panels with 15 layers of glass fiber mats were cured under microwave irradiation (Shull et al., 2000). The final panels (approximately 1.5 cm thick) were placed perpendicularly between a microwave source (1 W at 9 GHz), and receiver was used to monitor the microwave energy absorption by the composite during the cure cycle. Temperature profiles at various locations across the surface were probed using thermocouples so that the resin cure temperature data could be collected during microwave processing. Using that temperature information, the potential for localized microwave-accelerated cure to reduce the occurrence of material degradation from resin over-temperature was evaluated. Consequently, it was demonstrated that the application of microwave-assisted cure techniques reduced material degradation and residual stress in the composite. In addition, to elucidate the influence of microwave irradiation on the temperature profiles, a theoretical model was presented.

In turn, it was found that the morphology of a syntactic foam (hollow microspheres in a polymeric matrix) could be differently affected when cured under thermal and microwave conditions (Palumbo et al., 1998). The prepolymer used in this study was diglycidyl ether of bisphenol A (DGEBA) (DER 332) hardened with DDM. The glass microspheres (30% by weight) were given no special surface treatment and had wall thicknesses between 1 and 3 µm, with particle diameters ranging from 50 to 70 µm. The samples cured thermally were placed in an oven at 60 °C for 24 hours. The microwave cure was conducted in a microwave circuit (TE_{01} mode) with 2.45-GHz frequency during a three-step cycle: preheating 2 hours at 20 W, curing 1 hour at 25 W, and curing 1 hour at 30 W. Under microwave irradiation, due to faster energy transfer, more efficient crosslinking effects were observed at the interface, leading to higher intermodular crosslink density at the particle interface. As a consequence, slight but observable differences of mechanical behavior arise when curing the hollow-glass microsphere-filled epoxy resin composites in thermal and microwave conditions. The microwave-cured syntactic composites were found to be less ductile and a little more rigid due to a greater homogeneity (Palumbo et al., 1998).

Presently, commonly used marine yachts are built by manual lay-up with disadvantages of poor quality and too-long production time. Preliminary studies on the microwave curing of a polyester resin and the related composite material used in marine industry were presented and have shown that the microwave curing was a valid alternative method for the faster processing of laminated materials for structural applications (Bonnacorsi et al., 2005). Samples of polyester resin with 2% catalyst have been exposed to

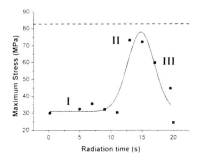
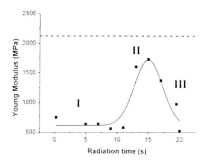

Fig. 5.3. Stress for pure resin and MAT 600 composite laminate cured by microwave. Reprinted from Bonnacorsi, L., Calabrese, L., Proverbio, E., Visco, A. 2005. 10th International Conference on Microwave and RF Heating, 302, with permission.

microwaves in a 2.45-GHz oven, at varying exposure time with a power output of 1,800 W. After that, laminates (10 cm × 10 cm) made with three layers of fiberglass and polyester resin have been prepared by manual lay-up and cured under microwave irradiation.

In order to better elucidate the effect of microwave irradiation, the flexural tests were performed on both the pure resin and the laminate composite at varying radiation times. The results of flexural stress are shown in Figure 5.3.

The mechanical properties of resins were not influenced by the microwave treatment when the resins were heated less than 11 to 12 seconds because the energy is sufficient only to activate the curing process and the resins were not fully cured (zone I). After this time, the resins quickly solidified and their maximum stress values were close to the horizontal dotted line, which represents the mechanical property of the resin cured at room temperature for 10 days (zone II). Over a curing time of radiation longer than 15 seconds, a rapid cure process was caused, which resulted in the appearance of ripples and bubbles in the sample (zone III). Finally, a long radiation time of the resin showed negative effects with a rapid decrease in the mechanical properties reaching that of the not fully cured material. Moreover, it was found that the use of high-power microwave irradiation during a short time induces the fast crosslinking and generates residual stress in the cured resin, which can cause matrix cracking.

Processing of neat unsaturated polyester resins (diallyl phthalate diluted with vinyltoluene) and composites with glass fiber under microwave irradiation was also studied (Hottong et al., 1991). As a microwave source, a single-mode resonant cavity operating at 2.45 GHz in TE_{111} mode with constant input power of 60 W was used. An allyl phthalate polyester was chosen for the neat resin studies, vinyltoluene (30 wt%) was used as the crosslinking monomer, and benzoyl peroxide (1%) was added as a initiator. Glass fiber (70 wt%) composites with diallyl phthalate polyester as matrix material were used in the composites investigation, and the prepreg was in the form of 1.6-cm-wide and 0.3-cm-thick tape. A thin film technique was applied in the reaction kinetics study to avoid large temperature gradients. Samples prepared on KBr disks were isothermally microwave cured at 85, 100, and 115 °C, while the extent of cure of the samples was monitored by means of Fourier transform-infrared spectroscopy. It was found that microwaves could initiate the reaction at a lower bulk temperature and shorter time than thermal heating. As the result, higher reaction rates were observed in microwave curing as compared to thermal curing. At lower polymerization temperatures, such as 85 °C, the ultimate extent of cure was higher under microwave than conventional thermal conditions.

Microwave heating was also used to enhance the manufacture of glass fiber–reinforced composites by pultrusion (Methven, 1999). The main benefits of MAP (microwave-assisted pultrusion) over conventional process include faster line speeds, reduced pulling forces, greater uniformity of cure, and reduced floor area in the case of simple profiles. Microwave heating may be used to preheat a pultrusion precursor upstream of the main pultrusion die, or it may be used in conjunction with a microwave transparent die as a direct replacement.

All MAP processes involve the use of a specific applicator. For the simple cylindrical profile, the preferred applicator is a single-mode (TM_{010}) cylindrical cavity (Fig. 5.4). This applicator has been used to manufacture solid cylindrical profiles based on glass fiber and a number of resins, including unsaturated polyester, urethane, acrylate, vinyl ester, epoxy, and phenolic.

The noncontacting nature of microwave heating is exploited in a "dieless" pultrusion process based on a traveling wave applicator (TWA) shown in Figure 5.5.

The waveguide was terminated by mitered bends that allow power in and excess power to be dumped into a water load. A hollow cylindrical tube welded to each bend prevented radiation leakage and at the same time allowed passage of the material to be heated through the TWA (Methven, 1999).

Microwave irradiation was investigated to cure and process nadic–end capped polyimide precursor (RP-46 resin) and glass/graphite/RP-46 composites (Liu et al., 1999). Processing of thick sections by conventional thermal process requires slow ramp rates and a long processing time. Therefore, the composite materials containing conducting fibers could be heated by the microwave process to achieve inside-to-outside heating patterns and quick heat ramp rates. Additionally, microwave process may enhance the bonding strength between resin and fiber matrix.

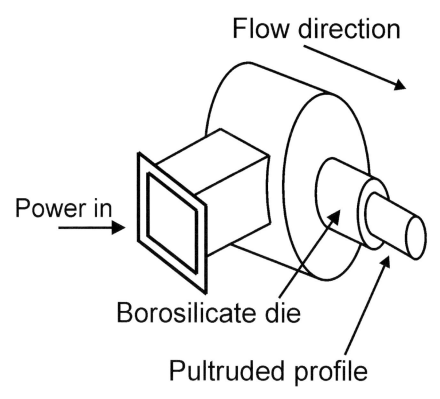

Flow direction

Power in

Borosilicate die

Pultruded profile

Fig. 5.4. Single mode cavity. Reprinted from Methven, J. 1999. Mater. Tech. Adv. Perf. Mater. 14:183, with permission.

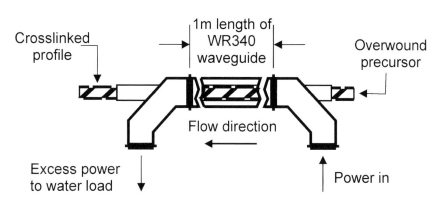

Crosslinked profile

1m length of WR340 waveguide

Overwound precursor

Flow direction

Excess power to water load

Power in

Fig. 5.5. MAP in a traveling wave applicator. Reprinted from Methven, J. 1999. Mater. Tech. Adv. Perf. Mater. 14:183, permission.

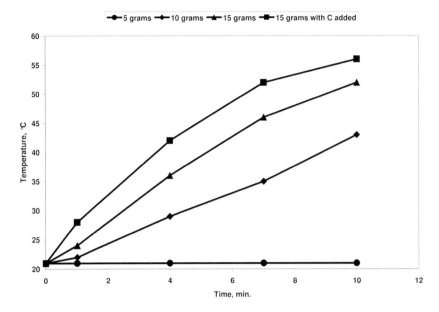

Fig. 5.6.   The effect of sample size on the microwave absorption of undried RP-46 resin. Reprinted from Liu, X.Q., Wang, Y.S., Zhu, J.H. 2004. J. Appl. Polym. Sci. 94:994, with permission.

Both neat resin and composite with glass and graphite cloth were obtained, and the effects of various parameters such as power level, mold material, and pressure were studied, using a microwave oven at a frequency of 2.45 GHz. It was shown that the sample size and geometry were important factors in microwave processes. For example, changing the sample size from 5 to 15 g caused a temperature increase of 32 °C in 10 minutes at the same power level as shown in Figure 5.6. Essentially, no coupling occurred between a sample of 5 g of polyimide resin and microwave energy, proving that a critical mass was required to absorb the microwave energy with a high efficiency.

Depending on the conditions, microwave cure of glass and glass/ graphite hybrid composites was accomplished in 0.6 to 2.16 hours, and the imidization of neat resin and composites was complete. Resin specimens containing only 0.057 wt% chopped graphite fibers resulted in complete imidization in 6 minutes. Glass and glass/graphite composites were fabricated by microwave irradiation with flexural strength and moduli equivalent from 50% to 80% of the properties of composites fabricated by conventional thermal processes (Liu et al., 1999).

In the next step, the application of microwave irradiation to the processing of carbon fiber–reinforced phenylethynyl-terminated polyimide compo-

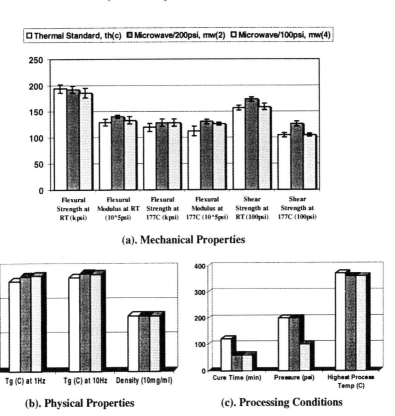

Fig. 5.7. Comparison of processing conditions and properties between microwave and thermally cured carbon fiber–reinforced phenylethynyl-terminated polyimide composites (PETI-5/IM7) composites. Reprinted from Fang, X., Scola, D.A. 1999. J. Polym. Sci. Part A Polym. Chem. 37:4616, with permission.

sites (PETI-5/IM7) was investigated and evaluated. Six different microwave cure cycles and three thermal processes were investigated. It was found that the polyimide prepreg showed no obvious difference in coupling with microwaves within a range of 2.4 to 7.0 GHz. Higher glass transition temperatures were observed in the microwave-cured composites. Thermally cured composites, fabricated from the same time-temperature cure cycles as the microwave processes, showed incomplete cure and much lower glass transition temperatures. Compared with the standard thermally cured composites, microwave-cured samples exhibited higher flexural strengths, moduli, and shear strength values. A microwave process was demonstrated that fabricated unidirectional polyimide-(carbon fiber) composites with superior thermal and mechanical properties relative to the thermal process in half the time required for the thermal process (Fig. 5.7) (Fang et al., 1999).

Nanocomposites

Nanocomposites based on polymer-clay systems have considerable interest for the development of new structural and functional materials. Poly(ethylene oxide) (PEO)-based nanocomposites with monntmorillonite, hectorite, and laponite were prepared by a melting intercalation procedure induced by microwave irradiation (Aranda et al., 2003). The microwave method takes advantage of smectic clays to absorb microwave energy, which activates water molecules in the hydration shell of the clay interlayer cations. The activation provokes the heating of the system, allowing the rapid intercalation of molecular organic compounds.

In the case of PEO/montmorillonite composite, the heating is mainly associated with the rotation of water molecules, which have a tendency to evaporate and open the clay layers. This phenomenon allows the entry of PEO chains and coordination of the interlayer cations (Fig. 5.8).

The intercalation compounds obtained using the microwave method showed an increase in ionic conductivity in comparison with those of materials prepared by conventional absorption methods. The samples of the nanocomposites were typically prepared using approximately 100 mg of a mixture of homoionic montmorillonite and PEO. The mixture was treated by microwave irradiation in a domestic oven working with maximum microwave power of 700 W. Variables such as the time of irradiation, microwave power, total amount of the mixture, and relative ratio of the components

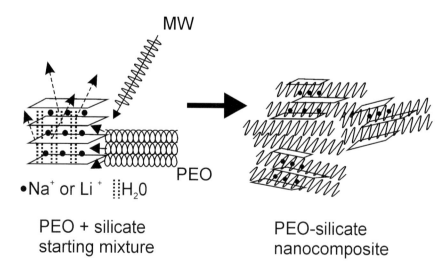

Fig. 5.8. Scheme of the microwave assisted melt-intercalation reaction. Reprinted from Aranda, P., Mosqueda, Y., Perez-Cappe, E., Ruiz-Hitzky, E. 2003. J. Polym. Sci. Part B Polym. Phys. 41:3249, with permission.

Table 5.3. Characteristic of PEO/sodium montmorillonte (MNA) prepared by microwave-assisted melting intercalation under different experimental conditions.

Variable	Experimental conditions	Variable range	Salient features in the nanocomposites
Time of irradiation	20:100 PEO/MNA	3–30 min	For t < 1-min, no intercalation
	MW power = 525 W		
	RH ≈ 55%		

Reprinted from Aranda, P., Mosqueda, Y., Perez-Cappe, E., Ruiz-Hitzky, E. 2003. J. Polym. Sci. Part B: Polymer Physics 41:3249, with permission.

were examined. The properties of PEO/sodium montmorillonite nanocomposites prepared under different conditions are shown in Table 5.3.

It was found that intercalations were not produced for irradiation times shorter than 5 minutes. In the case of longer irradiation times, the XRD patterns confirmed the PEO intercalation, showing interlayer spacing of around 1.8 nm and different degrees of disorder in the stacking of the clay particles depending on the time of irradiation (Fig. 5.9). Longer periods of irradiation gave materials with better stacking order.

A few experiments were carried out in a microwave reactor that focused the irradiation directly on the sample, allowed the control of reaction temperature, and kept a constant temperature just slightly above the melting point of PEO by the adjustment of power. In such conditions, the water content and homogeneity of the starting mixture have to be very carefully controlled because high energy of the focused microwaves may produce locally high temperature, in which the polymer could be degraded.

A novel method for the preparation of semiconductor/polymer nanocomposites by microwave irradiation was reported (He et al., 2003). Microwave irradiation has been used to synthesize nanocomposites of polyvinylcarbazole with CdS in a few minutes without stirring, in which the polymerization and the formation of the crystalline inorganic salt nanoparticles progressed simultaneously.

The approach involved the preparation of solutions consisting of inorganic salts and organic monomer. Pyridine was chosen as a solvent due to its high permanent dipole moment and because it can be used as a susceptor of the microwave irradiation to promote the N-vinylcarbazole (NVK) polymerization and formation of CdS nanoparticles. In a typical procedure, the reaction solution was prepared by dissolving 0.5 g of NVK, 0.005 g of AIBN, 0.039 g of tiourea, and 0.071 g of cadmium acetate in 5 mL of pyridine. The reaction mixture was pumped with pure nitrogen for approxi-

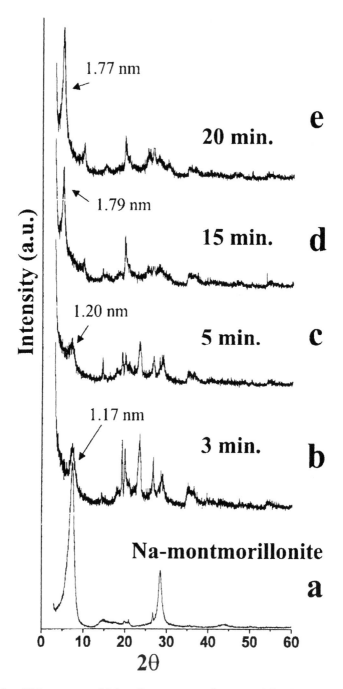

Fig. 5.9. XRD patterns of (**a**) sodium montmorillonite and (**b–e**) PEO/sodium montmorillonite nanocomposites prepared by the microwave irradiation treatment (525 W) of 20:100 PEO/sodium montmorillonite mixtures for the times given. Reprinted from Aranda, P., Mosqueda, Y., Perez-Cappe, E., Ruiz-Hitzky, E. 2003. J. Polym. Sci. Part B: Polymer Physics, 41:3249, with permission.

mately 20 minutes. The vials were irradiated in the center of a domestic microwave oven at microwave irradiation of 300 W for 5 minutes.

This one-step method not only shared advantages of energy saving but also successfully developed a convenient method for fabrication of nanocomposites, in which crystalline inorganic nanoparticles are homogeneously dispersed in polymer matrix.

Effects of Different Fillers

Microwave heating has a number of advantages in comparison to the conventional method due to the ability to heat a part of composite material directly through specific interaction of electromagnetic radiation with selected types of materials. Therefore, the various techniques that use microwave irradiation are employed for forming, welding, joining, and bonding polymers and composites. Most thermoplastics are relatively transparent for microwave irradiation and they do not absorb microwaves to a sufficient extent to be heated. Enhanced microwave heating can result from the use of fillers such as carbon black, ferrites, or conducting polymers and chiral microinclusions.

The application of inherently conducting polymers (ICPs) was also reported for welding thermoplastics (Kathirgamanathan, 1993). However, the use of ICPs as microwave absorbers for welding was suggested earlier (Epstein et al., 1991). In a typical experiment, a small amount of ICP as either powder (20 mg) or type (two strips with dimensions of 20 mm × 2 mm × 40 μm were placed at the upper surface of the plastic [10 cm × 2 cm × 0.1 cm]). The second piece was placed on the top and clamped. The clamp assembly was then placed in a household microwave oven with a total power of 500 W for 2 to 120 seconds; the results are summarized in Table 5.4. The method can be applied to any thermoplastics (Kathirgamanathan, 1993).

A polymer with a silicone backbone structure and chiral side group was added to the ABS powder (Varadan et al., 1991). According to this investigation, chirality helped to localize microwave heating when a suitable applicator was used, and the addition of chiral material to fiber-reinforced thermoset or thermoplastic composites enhanced the absorption and thermal conductivity of the basic polymer and brought these properties to the same level as fibers like graphite or Kevlar under microwave exposure.

Different fillers such as talc, zinc oxide, and carbon black are used to increase the susceptibility of common polymers to microwave processing (Harper et al., 2005). In this investigation, a modified 800-W domestic microwave oven was used to heat materials. The relative temperature rise for various susceptors at the same volume percent loading for exposure to 200

Table 5.4. Microwave welding of thermoplastics using inherently conducting polymers.

Substrate to welded	Conducting polymer[a]	Conductivity (S cm^{-1})	Time to weld (s)
Polycarbonate	PPPTS(p)	22.0 ± 2.0	100 ± 10
Polycarbonate	PAPTS(p)	9.0 ± 1.0	120 ± 5
Polycarbonate	A(t)	1.0 ± 0.1	5 ± 1
Polycarbonate	B(t)	12.0 ± 3.0	2 ± 1
Polycarbonate	C(t)	(5.7 ± 0.3) × 10^{-2}	40 ± 2
Polypropylene	A(t)	1.0 ± 0.1	5 ± 1
Polypropylene	B(t)	12.0 ± 3.0	2 ± 1
Polypropylene	C(t)	(5.7 ± 0.3) × 10^{-2}	60 ± 5
Polyethylene	PPPTS(p)	22.0 ± 2.0	5 ± 2
Polyethylene	PAPTS(p)	9.0 ± 1.0	15 ± 2
Polyethylene	A(t)	1.0 ± 0.1	30 ± 5
Polyethylene	B(t)	12.0 ± 3.0	2 ± 1

Reprinted from Kathirgamanathan, P. 1993. Polymer 34:3105, with permission.
[a] PPPTS, polypyrrole *p*-toluenesulfonate; PAPTS, polyaniline *p*-toluenesulfonate; p, powder (thickness of powder on the weld 100 μm); t, tape, A(t), nonwoven polyester tape impregnated with PPPTS (thickness 110 μm), B(t), microporous polyethylene impregnated with PPPTS (thickness 40 μm), C(t), nonwoven polyester impregnated with carbon black (110 μm).

W of microwave power for 60 seconds is shown in Table 5.5. As was expected, a test on unfilled HDPE under the same conditions gave no significant temperature rise.

It was clear from preliminary screening that carbon black is the most effective filler for imparting microwave heat ability to HDPE and the efficiency was directly proportional to its surface area (Fig. 5.10).

More extensive measurements were carried out on carbon black of different grades. Relative temperature rises and heat efficiency that represents

Table 5.5. Relative temperature rise for HDPE at 16% v/v loading of different fillers.

Filler	Temperature rise (°C)
Carbon black N550	177 ± 10
Zinc oxide	30 ± 5
Talc	3 ± 1

Reprinted from Harper, J.F., Price, D.M. 2005. 10th International Conference on Microwave and RF Heating, 298, with permission.

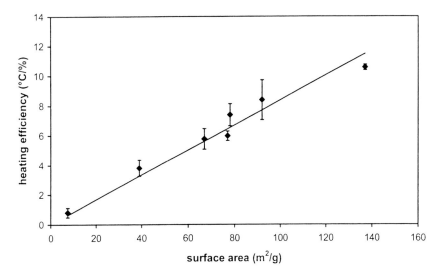

Fig. 5.10. Relationship between microwave heating efficiency and surface area for carbon black. Reprinted from Harper, J.F., Price, D.M. 2005. 10th International Conference on Microwave and RF Heating, 298, with permission.

the slope of a linear regression of temperature rise versus carbon black content are included in Table 5.6.

Carbon black had a detrimental effect on the mechanical properties of the HDPE for the content above 10% to 15%, and therefore, in such cases,

Table 5.6. Relative temperature rise for HDPE filled with different carbon blacks.

Product code	Surface area (m²/g)	Oil absorbtion (cc/100 g)	$\Delta T/°C$ at different carbon black loading (% w/w)				Heating efficiency (°C/%)
			5	10	20	30	
MT	7.5	41	1	1	3	22	0.8 ± 0.3
N550	39	121	3	5	62	90	3.8 ± 0.6
C7055	67	166	3	10	91	137	5.8 ± 0.7
N330	77	102	4	34	85	156	6.0 ± 0.3
N326	78	72	4	21	92	186	7.4 ± 0.7
N339	92	120	4	23	87	215	8.3 ± 1.3
N115	137	113	16	66	174	>400	10.5 ± 0.2

Reprinted from Harper, J.F., Price, D.M. 2005. 10th International Conference on Microwave and RF Heating, 298, with permission.

selecting a grade with a high surface area is recommended (Harper et al., 2005).

A phenomenological kinetic model was used to describe the autocatalytic behavior of curing reaction of both neat and carbon-doped epoxy system (DGEBA/DDS) (Zhou et al., 2003). In this work, a new approach was used to study the mechanism of microwave heating. Carbon black was added to epoxy to modify the microwave thermal effect without significantly affecting the nonthermal alignment of polar groups in the electromagnetic field. The effect of carbon on microwave absorption by polymer reactants is hypothesized in Figure 5.11.

In microwave curing of the neat resin, microwaves are absorbed by the functional groups and dissipated into heat and then the heat is transferred to the entire molecules. Therefore, the local temperature should be higher than that of the bulk temperature. When carbon black is filled into the resin, much fewer microwaves are absorbed by the functional groups because most microwaves are absorbed by carbon owing to its much better dielectric properties. Heat is then transferred from carbon to the resin molecules and the functional groups of the resins.

Microwave and thermal curing of the neat and doped resin was performed at 145, 165, and 185 °C. The extent of cure of the resin was tested with DSC as a function of curing time. A phenomenological kinetic model was used to describe the epoxy autocatalyzed curing reaction for a stoichiometric reactant mixture (Eq. 5.1):

$$d\alpha/dt = (k_1 + k_2\alpha^m)(1 - \alpha)^n \qquad (5.1)$$

where α is the extent of cure, t is time, k_1 is the noncatalytic polymerization reaction rate constant, k_2 is the autocatalytic polymerization reaction rate constant, m is the autocatalyzed polymerization reaction order, and n is the noncatalyzed polymerization reaction order. This autocatalyzed reaction model has been widely used to represent adequately the cure kinetics of thermosetting resins. In both thermal and microwave curing, no obvious trend was found for the reaction order constants m and n as functions of carbon concentration. The rate constants k_1 and k_2 for the thermal and microwave curing are shown in Figures 5.12 and 5.13, respectively.

In microwave curing of the resin with different carbon concentration, the reaction rate constants decreased with increasing carbon concentration. However, it was hypothesized that the presence of carbon black weakened the localized microwave superheating effect, because the conducting carbon absorbed more microwaves during the curing process (Zhou et al., 2003)

Intrinsically conductive polyaniline (PANI) composite gaskets were used to microwave (2.45 GHz) weld high-density polyethylene (HDPE) bars (Wu

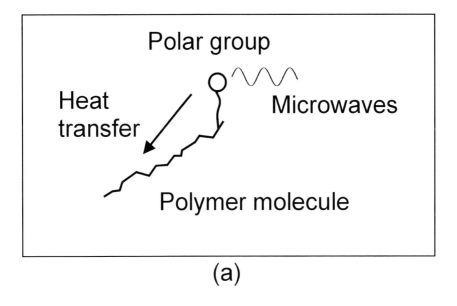

(a)

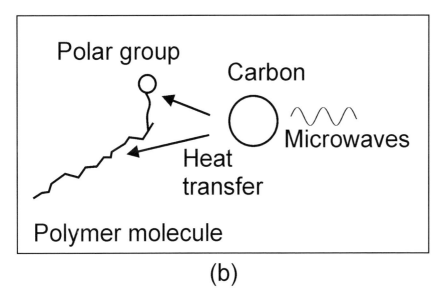

(b)

Fig. 5.11. Microwave interaction with (**a**) neat epoxy and (**b**) epoxy doped with carbon. Reprinted from Zhou, S., Hawley, M.C. 2003. Composite Struct. 61:303, with permission.

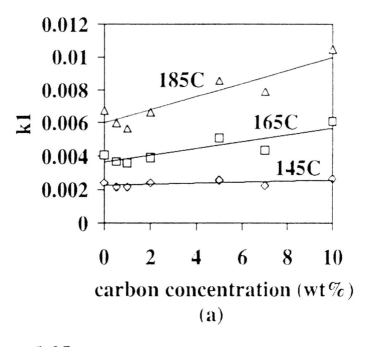

(a)

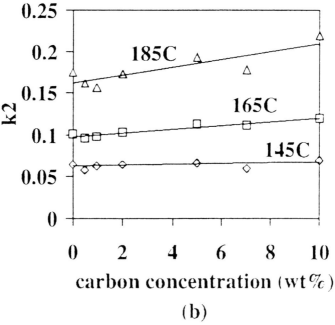

(b)

Fig. 5.12. Effect of carbon concentration on (a) noncatalytic reaction rate and (b) catalytic reaction rate in thermal curing. Reprinted from Zhou, S., Hawley, M.C. 2003. Composite Struct. 61:303, with permission.

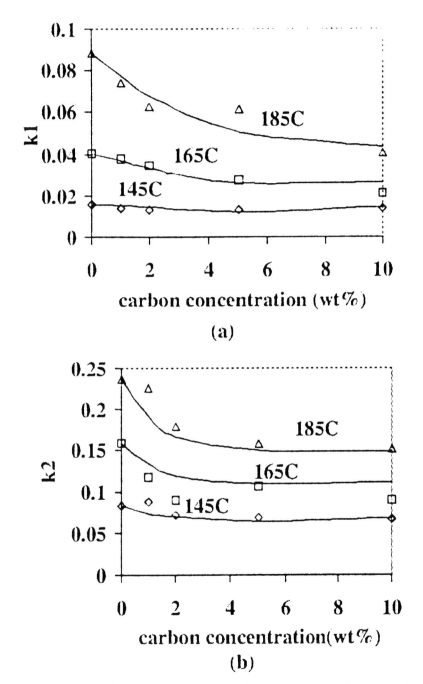

Fig. 5.13. Effect of carbon concentration on (**a**) noncatalytic reaction rate and (**b**) catalytic reaction rate in microwave curing. Reprinted from Zhou, S., Hawley, M.C. 2003. Composite Struct. 61:303, with permission.

et al., 1997). Two composite gaskets were made from a mixture of HDPE and PANI (50 and 60 wt%) powders in different proportions. The mixtures were compression molded in a hot (180 °C) press. Adiabatic heating experiments were used to estimate the internal heat generation. It was observed that the temperature rise rate (heat generation rate) decreases with increasing temperature. The decrease in heat generation rate was probably due to a reduction in the electrical conductivity of PANI at elevated temperatures. During microwave welding, increase in the heating time resulted in the development of a thicker molten layer in the parts, which improved the joint strength. The maximum tensile joint strength was achieved using a 60 wt% PANI gasket with a heating time of 60 seconds and a welding pressure of 0.9 MPa. This resulted in tensile weld strength of about 25 MPa, which equals the tensile strength of the bulk HDPE (Wu et al., 1997).

Microwave irradiation was examined as an alternative to conventional heating for joining composite structures (Thostenson et al., 1999). Through proper material selection, microwaves are able to penetrate the substrate materials and cure the adhesives in situ. Selective heating with microwaves is achieved by incorporating interlayer materials that have high dielectric loss properties relative to the substrate materials. A processing window for elevated temperature curing of an epoxy paste adhesive system was developed and composite joint systems were manufactured using conventional and microwave techniques and tested in shear. The adhesive joint systems were cured at room temperature under vacuum pressure for 24 hours. Joint systems were allowed to cure: C1—at room temperature for 10 days, C2—at room temperature for 5 months, and C3—at 90 °C for 1 hour. Another joint system (M1) was processed using microwaves (2.45 GHz, 6 kW) and heated at a rapid heating of 20 °C/min to a final temperature of 90 °C. Microwave curing resulted in enhanced shear strength samples M1 (10% higher than C1, 70% higher than C3) and less scatter experimental data (Thostenson et al., 1999).

Ku and others (2000) applied variable-frequency microwave (VFM) facilities (2 to 18 GHz) for joining parts made of thermoplastic composites and found that a selective heating over a large volume at a high-energy coupling efficiency can be obtained. Five different thermoplastic polymer matrix composites were processed, including 33 wt% random carbon fiber (CF)– or glass fiber (GF)–reinforced polystyrene (PS-CF, PS-GF), low-density polyethylene (LDPE-CF, LDPE-GF), and polyamide 66-GF. Bond strengths of lap joints of the composites were tested in shear, and the results were compared with those obtained using fixed frequency (2.45 GHz). The coupling agent used was a two-part adhesive (i.e., 100% liquid epoxy resins and 8% amine hardener). Considering the quality of bond brought about by processing using different microwaves facilities, hot spots were found on the joint of LDPE-CF bound using a fixed frequency of 2.5 GHz.

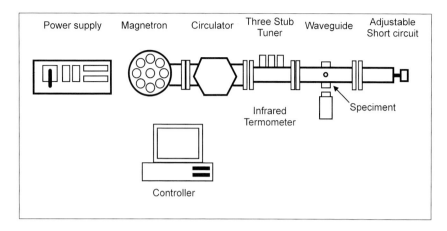

Fig. 5.14. Schematic representation of the set-up for microwave curing of adhesive joints. Reprinted from Olofinjana, A., Yarlagadda, P.K.D.V., Oloyede, A. 2001. International Journal of Machine Tools & Manufacture 41:209, with permission.

The joint of PS-CF processed by variable frequency was perfect. It could therefore be argued that VFM produces stronger bonds for the two materials, PS and LDPE, with excellent quality of joint properties (Ku et al., 2001).

In a number of cases, microwave curing can improve the mechanical properties of epoxies, but a precise temperature control is important because inadequate control can lead to localized material damage (Olofinjana et al., 2001). An electromagnetic microwave generator with variable power output of 0 to 1,950 W at a frequency of 2.45 GHz was used for curing the two-component epoxy adhesives (rapid araldite, RS high strength) between three types of adherents (PMMA, ABS, and UHMWPE). Schematic representation of the set-up for microwave curing of adhesive joints is shown in Figure 5.14.

The magnetron was interfaced with a computer that allowed the microwave input power and exposure time for various tests to be preset and monitored. An infrared thermometer with an accuracy of ±1 °C was used to facilitate the computer control. The results of the mechanical tests for samples processed at different microwave conditions are presented in Table 5.7.

In all the cases, significant increases in bond strength were observed, except for ABS-RS, where the effect of microwave irradiation was minimal. These results showed that a power threshold above which microwave heating ceases to be beneficial occurred in many investigated cases. Microwave curing based on temperature control produced a better bond strength and improved quality of adhesive joints compared to curing at ambient condi-

Table 5.7.   Bond strength (N/mm^2) of ambient and microwave cured samples.

Adhesive	Adherend	Mean peak bond strength, N/mm^2 at different curing conditions			
		ambient	300 W	450 W	600 W
Rapie	PMMA	2.51	5.50	4.00	4.85
Araldite	ABS	2.68	5.23	4.82	4.71
	UHMWPE	0.70	1.28	1.88	1.52
	PMMA	2.43	3.77	3.66	4.99
RS high strength	ABS	2.12	2.02	2.46	2.98
	UHMWPE	0.96	1.50	1.15	1.16

Reprinted from Olofinjana, A., Yarlagadda, P.K.D.V., Oloyede, A. 2001. International Journal of Machine Tools & Manufacture 41:209, with permission.

tions. The bond strength in general varied with microwave power attaining a maximum at 400 W and, for most samples, dropping at 600 W. A probable reason for such a behavior might be due to localized ordering of molecular aggregates and its associated embrittlement of the joints at this energy level.

In summary, a comparison of thermal and microwave cure assumes a new dimension when the temperature distribution inside the sample is considered, and that is where the scientific challenge lies. The fundamental difference in the heat transfer during reactive processing in thermal and microwave fields is that microwave energy, in contrast to thermal heating, is supplied directly to a large volume, thus avoiding the thermal lags associated with conduction and/or convection. Consequently, temperature gradients and the excessive heat build-up during thermal processing could be reduced by a microwave power control (Mijovic et al., 1998). In large polymer composite structures, high temperatures caused by exothermic resin cure can degrade the mechanical properties of the composites.

References

Aranda, P., Mosqueda, Y., Perez-Cappe, E., Ruiz-Hitzky, E. 2003. J. Polym. Sci. Part B: Polymer Physics 41:3249.

Bonnacorsi, L., Calabrese, L., Proverbio, E., Visco, A. 2005. 10th International Conference on Microwave and RF Heating, 302.

Epstein, A., MacDiarmid, A. 1991. In: Science and Application of Conducting Polymers. Salaneck, W.R., Clark, D.T. (eds), Adam Hilger, London.

Fang, X., Scola, D.A. 1999. J. Polym. Sci. Part A: Polym. Chem. 37:4616.

Forysth, A., Whittaker, A.K. 1999. J. Appl. Polym. Sci. 74:2917.

Harper, J.F., Price, D.M. 2005. 10th International Conference on Microwave and RF Heating, 298.

He, R., Qian, X., Yin, J., Bian, L., Xi, H., Zhu, Z. 2003. Mater. Lett. 57:1351.

Hottong, U., Wei, J., Dhulipala, R., Hawley, M.C. 1991. Proceedings of the 93rd Annual Meeting of the American Ceramic Society, 587.

Kathirgamanathan, P. 1993. Polymer 34:3105.

Ku, H.S., MacRobert, M., Siores, E., Ball, J.A.R. 2000. Plast. Rubber Compos. 29:278.

Ku, H.S., Siu, F., Siores, E., Ball, J.A.R., Blickblau, A.S. 2001. J. Mater. Process. Tech. 113:184.

Liu, X.Q., Wang, Y.S., Zhu, J.H. 2004. J. Appl. Polym. Sci. 94:994.

Liu, Y., Xiao, Y., Sun, X., Scola, D.A. (1999). J. Appl. Polym. Sci. 73:2391.

Methven, J. 1999. Mater. Tech. Adv. Perf. Mat. 14:183.

Mijovic, J., Corso, W.V., Nicolais, L., d'Ambrosio, G. 1998. Polym. Adv. Technol. 9:231.

Olofinjana, A., Yarlagadda, P.K.D.V., Oloyede, A. 2001. International Journal of Machine Tools & Manufacture 41:209.

Palumbo, M., Tempesti, E. 1998. Acta Polym. 49:482.

Shull, P.J., Hurley, D.H., Spicer, J.W.M., Spicer, J.B. 2000. Polym. Eng. Sci. 40:1157.

Thostenson, E.T., Chou, T.-W. 1997. Proc. 12th Annu. Meeting Amer. Soc. Composites, Dearborn, 931.

Thostenson, E.T., Chou, T.-W. 1998. Proc. 13th Annu. Meeting Amer. Soc. Composites, Baltimore.

Thostenson, E.T., Chou, T.-W. 1999. Adv. Aerospace Mater. Struct. 89.

Thostenson, E.T., Chou, T.-W. 2001. Polym. Composites 22:197.

Varadan, V.K., Varadan, V.V. 1991. Polym. Eng. Sci. 31:470.

Wu, C.-Y., Benatar, A. 1997. Polym. Eng. Sci. 37:738.

Zhou, S., Hawley, M.C. 2003. Composite Struct. 61:303.

6

RENEWABLE RESOURCES FOR THE PREPARATION OF POLYMERIC MATERIALS AND POLYMER MODIFICATION

Vegetable Oils

Vegetable oils are an excellent renewable source of raw materials for manufacturing polymer components with hydroxyl groups. The transformations of the double bonds of triglycerides to hydroxyl groups and their application for polymer synthesis have been the subject of a number of studies (Badri et al., 2001; Chian et al., 1998; Guo et al., 2000; Hofer, 1999; Hong et al., 2002; Prociak, 2006; Zlatanic et al., 2004). Taking into account that European countries have a vast potential regarding cultivation of oil plants, these investigations may result in a significant source of modern, sustainable, more ecological polymers. The real commercial development of natural oil–derived polyols (NOPs) in polymers is still on a small scale but an ever-growing number of users is expected. Vegetable oils are triglycerides (i.e., esters of different fatty acids and glycerol). The composition of the fatty acids contained in the vegetable oils determines the further use of the oils. The types of unsaturated fatty acids in triglycerides determine the distribution of hydroxyl groups in oil-based polyols. The number of hydroxyl group depends on the type of unsaturated fatty acids and the agent used for oxirane ring opening. Oils such as linseed or rapeseed and soybean can also be useful in polymer formulation, but, first, they need to be transformed. There are several methods of converting them into the raw materials contained the hydroxyl group capable of forming new polyurethanes. Through epoxidation, the double bonds of the triglycerides are transformed into oxirane rings. In the second step, the epoxidized oil can be converted into the polyols by ring-opening reaction with alcohols (i.e., methanol, ethylene glycol, propylene glycol, glycerol, etc.).

A two-step process was applied for the preparation of rapeseed and linseed oil based polyols (Prociak, 2006). In the first step, the double bonds of the triglycerides were transformed into oxirane rings through the epoxidation with acetate peroxyacid. In the second step, the epoxidized oils

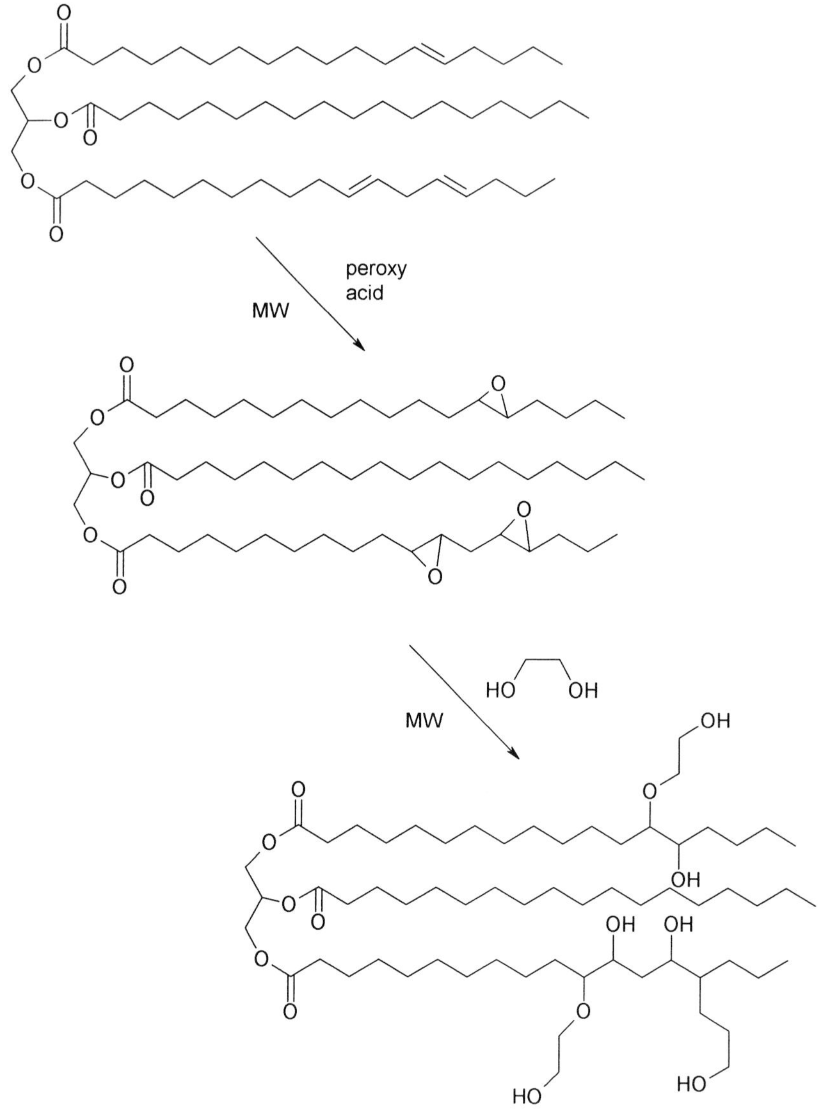

Fig. 6.1. Two-step process of preparation of oil-based polyols.

were converted into the polyols using monoethylene glycol or diethylene glycol (Fig. 6.1). In the case of the polyol based on rapeseed and linseed oil, both steps of the process (epoxidation and oxirane ring opening) were carried out under microwave irradiation. It allowed reduction in the reaction time of the epoxidation step by approximately 60% and of hydroxyla-

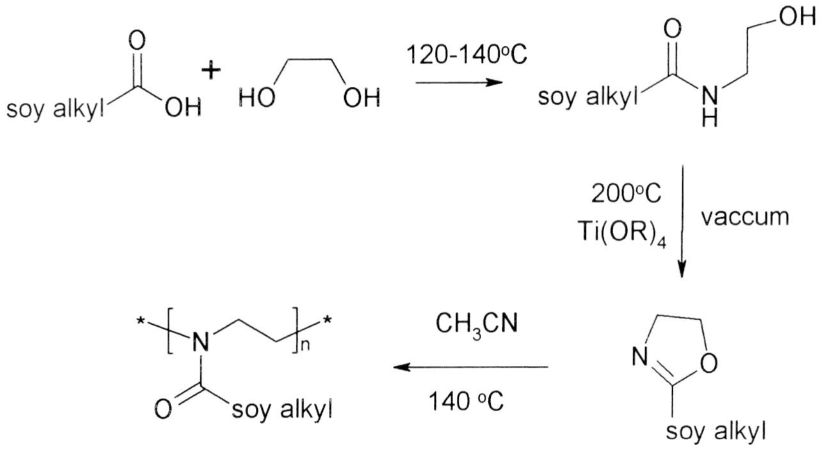

Fig. 6.2. Coupling soybean fatty acids with ethanolamine followed by ring closure.

tion step by approximately 75% in comparison to the same process run under conventional thermal conditions.

Lately, the oxirane ring-opening process of epoxidized rapeseed oils under microwave irradiation was investigated in higher scale (i.e., 660 g of reactants) (Mikolajek, 2006). Dietylene glycol was used in the molar ratio to oxirane rings of 1:1. The initial epoxy number of the epoxidized rapeseed oil was in the range of 0.28 to 0.30 mol/100 g. The reaction mixture was stirred by means of a magnetic stirrer. The process was carried out for 40 minutes while different powers of the microwave reactor were applied in order to keep the reaction temperature at approximately 105 to 110 °C.

It was found that the reaction time of 8 minutes under maximum power of 360 W was sufficient to reach a temperature of approximately 100 °C. Further microwave irradiation with power in the range of 0 to 150 W was needed to maintain temperatures of 105 to 110 °C. In such a way, the rapeseed oil–based polyol was prepared with primary and secondary hydroxyl groups capable of reacting with isocyanates. The final epoxy number was lower than 0.04 mol/100 g, and the hydroxyl number was approximately 260 mg KOH/g. The polyols were successfully applied in a two-component polyurethane system for the preparation of microporous elastomers.

Recently, the microwave-assisted polymerization of a sustainable soy-based 2-soy-2-oxazoline (SoyOx) monomer was reported (Hoogenboom et al., 2005). This monomer was synthesized by coupling soybean fatty acids with ethanolamine followed by ring closure in the presence of a titanium catalyst (Fig. 6.2). The resulting 2-soy-2-oxazoline is a mixture of 2-oxazolines with different fatty acid side chains. Furthermore, the unsatu-

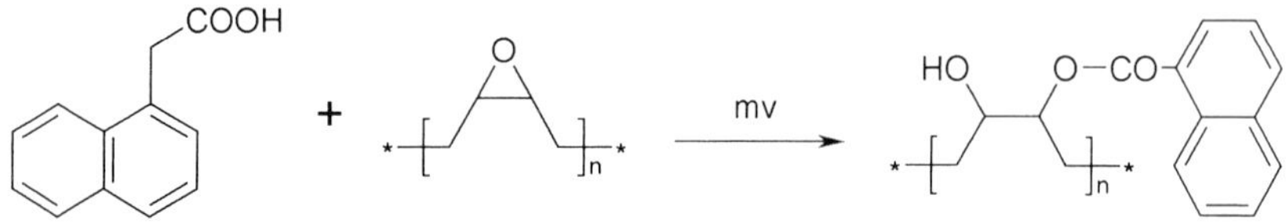

Fig. 6.3. Fixation of α-naphthyl acetic acid (NAA) onto epoxidized liquid natural rubber. Reprinted from Huy, H.T., Buu, T.N., Dung, T.T.K., Han, T.N., Qaung, P.V. 1996. J. Mater. Sci., Pure Appl. Chem. A33:1957, with permission.

rated nature of the fatty acids is transferred into the SoyOx monomer, providing crosslinking possibilities for the resulting polymers.

The microwave-assisted polymerization of the purified SoyOx was performed in acetonitrile with methyltosylate as initiator [(M)/(I) = 60] at 140 °C (superheated conditions), because these conditions were previously found to be the optimal conditions for the polymerization of 2-ethyl-2-oxazoline (Wiesbrock et al., 2004). The kinetics of the SoyOx polymerization were investigated by quenching the polymerization after different reaction times with water.

The $\ln[(M)0/(M)t]$ revealed a linear dependence on the reaction time, demonstrating a constant concentration of cationic polymer species indicative of a living polymerization. The microwave-assisted polymerization of SoyOx in acetonitrile reached full conversion within 15 minutes of polymerization time (Hoogenboom et al., 2005).

α-Naphthyl acetic acid (NAA) is a plant growth stimulator that is broadly used in agriculture. In order to prolong its activity on plants, some attempts have been made to fix NAA onto polymers. For this reason, the ring-opening reaction of epoxidized liquid natural rubber (ELNR) (Fig. 6.3) with naphthyl acetic acid under microwave irradiation was studied (Huy et al., 1996). In this case, the reactions were carried out in a chlorobenzene solution (1.2 mL) in a single-mode microwave reactor, in which the mixtures of naphthylacetic acid (2.5 to 10.2 mmol), ELNR (2.5 to 10.2 mmol), and TEBA (0.083 to 0.34 mmol) were irradiated for 5 to 15 minutes. The reaction mixture was precipitated into methanol, while the conversion of the fixation was calculated based on titration of residual NAA.

The dependence of the conversion on the reaction time is presented in Table 6.1. The maximum conversion (i.e., 44%) of epoxy groups under microwave irradiation at 110 °C was reached after 10 minutes, whereas under similar conventional conditions the conversion reached 33% after 24 hours of heating.

Fatty acid methyl esters (FAMEs) are obtained with glycerol by transesterification of natural triglycerides (Fig. 6.4). They are important reactants

Table 6.1. Dependence of the conversion on the reaction time.

Reaction time, t (min)	Conversion, %	Temp. at the end[a] (°C)
05	33	109
08	35	109
10	44	111
15	44	111

Note. ELNR 38 = 2.5 mmol, NAA = 2.5 mmol, TEBA = 0.083 mmol, chlorobenzene = 1.2 mL, P = 150 W.

[a] Temperature was determined at the end of the reaction by thermometer.

Reprinted from Huy, H.T., Buu, T.N., Dung, T.T.K., Han, T.N., Qaung, P.V. 1996.
J. Mater. Sci., Pure Appl. Chem. A33:1957, with permission.

in a number of processes and can be used to obtain products for food additives, cosmetics, detergents, pharmaceuticals, polymers, and alternative
diesel oil (Biodiesel).

For this reason, transesterification reactions were tested under microwave
conditions with rapeseed oil and methanol in the presence of homogeneous-
heterogeneous basic catalysts such as $Ba(OH)_2$-H_2O and heterogeneous

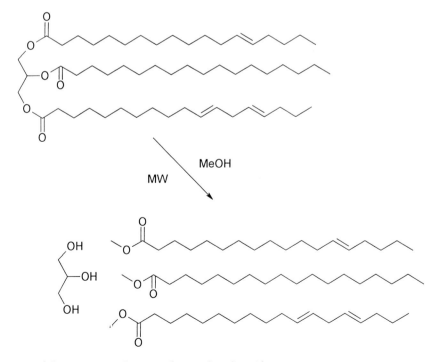

Fig. 6.4. Transesterification of natural triglycerides.

Table 6.2. Transesterification tests using Ba(OH)$_2$ • H$_2$O with different methanol/oil mol ratio at atmospheric pressure and 65 °C.

Reaction time (min)	Heating	FAMEs, yield (%)	
		Methanol/oil mol ratio 9:1	Methanol/oil mol ratio 18:1
5	MW	83.9	94.9
10	MW	97.8	100
15	MW	100	—
5	Δ	80.2	88.9
10	Δ	95.7	96.8
15	Δ	97.2	97.9
30	Δ	100	100

Reprinted from Mazzocchia, C., Kaddouri, A., Modico, G., Nannicini, R., Martini, F., Marengo, S. 2005. Fatty Acid Methyl Esters Synthesis from Triglycerides over Homogenous/Heterogeneous Catalysts in the Presence of Microwaves. Proceedings of 10th International Conference on Microwave and RF Heating. Modena, Italy, with permission.

acidic catalysts such as K10 and KSF montmorillonites (Mazzocchia et al., 2005). The reactions under microwave conditions were performed within a multimode microwave reactor with a power up to 1,000 W. The tests at atmospheric pressure under conventional and microwave conditions were carried out in Pyrex vessels, whereas under pressurized conditions (i.e., 20 bar), the reactions were run in a Buchi autoclave and Teflon vessels, respectively. It was found that microwave irradiation allows high conversions up to 100% of FAMEs from triglycerides in a few minutes while Ba(OH)$_2$-H$_2$O was employed as a catalyst (Table 6.2). With other catalysts, the conversions were much lower; moreover, in the presence of K10 and KSF montmorillonites, the glycerol that was obtained was brown and dark.

Transesterification reaction of rapeseed oil under microwave conditions with methanol was also tested in the presence of sulfuric acid as a catalyst (Bogdal et al., 2006). For this purpose, 100 g of rapeseed oil, 23.7 g of methanol, and 1.6 mL of sulfuric acid were placed in 250-mL round-bottomed flask equipped with an upright condenser and magnetic bar, irradiated, and stirred in a multimode microwave reactor for 30 minutes at 65 to 68 °C (reflux). After cooling to room temperature, the reaction mixture was separated into FAMEs and glycerol phases with high conversion of triglycerides. Then the double bonds of FAMEs were epoxidized and converted into polyols by the ring opening reaction of oxirane rings using monoethylene glycol (MEG) or diethylene glycol (DEG) as was described for triglycerides at the beginning of this chapter.

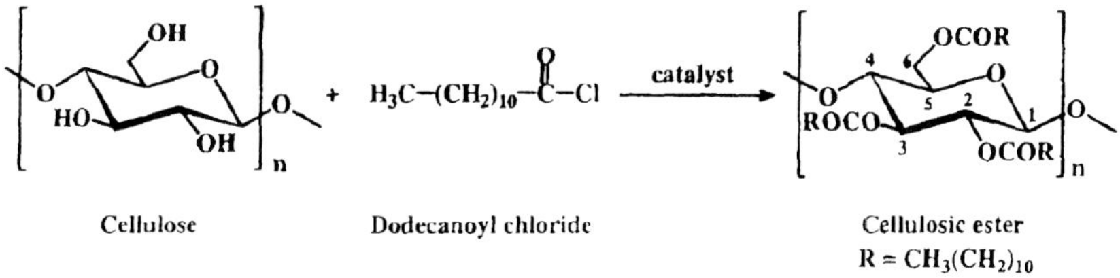

Fig. 6.5. Esterification of cellulose with dodecanoyl chloride under microwave conditions. Reprinted from Gourson, C., Benhaddou, R., Granet, R., Krausz, P., Saulnier, L., Thibault, J.F. 1999. C. R. Acad. Sci. Paris 2:75, with permission.

Cellulose

Cellulose is the most abundant polymer on Earth and accounts for 40% to 50% of the mass of wood. Long-chain aliphatic esters of cellulose were identified as potential biodegradable plastics, and the discovery of non-degrading solvents for cellulose such as the lithium chloride and N,N-dimethylacetamide (LiCl/DMAc) increases the interest in the modification because homogeneous reaction conditions present many advantages (Heinze et al., 1998), as follows:

- Allow obtaining cellulose derivatives with well-controlled degree of substitution
- Substitution pattern is regular along the polymer backbone
- Reactions can be carried out in the reduced time scale

First, it was demonstrated that applying dodecanoyl chloride under microwave conditions the esterification of cellulose can be achieved within 6 to 9 minutes (Fig. 6.5) (Gourson et al., 1999). In a general procedure, a mixture of cellulose, acid chloride, and catalyst (4-dimethylaminopyridine) was irradiated in an open vessel placed in a domestic microwave oven. Prior to the reaction, the homogeneity of the mixture was ensured by sonification.

The best results under microwave irradiation were obtained when the reaction mixture was absorbed on inorganic support such as K_2CO_3 or Al_2O_3, which allow obtaining yields in the range of 17% to 30% with 2.1 to 2.5 degrees of substitution (i.e., the number of grafted dodecanoyl residues per anhydroglucose unit). It was shown that within the same reaction time, under similar conventional conditions, no reaction occurred.

Then it was shown that long-chain cellulose esters can be synthesized by acylation of cellulose in an LiCl/DMAc solution with lauroyl chloride in the presence of N,N-dimethyl-4-aminopyridine (DMAP) as a catalyst (Satge et

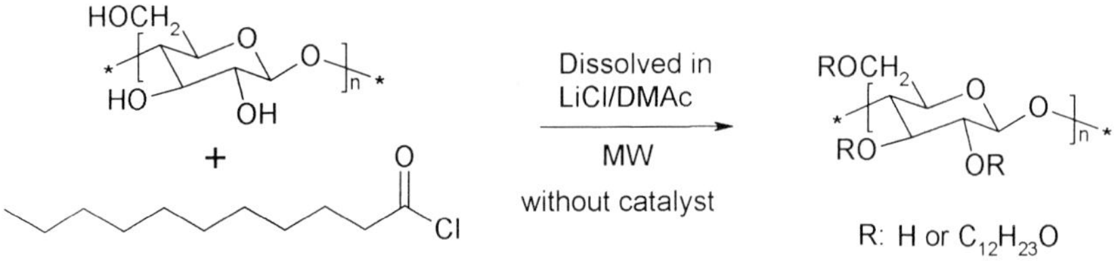

Fig. 6.6.   Microwave-assisted acylation of cellulose by chlorides of fatty acids.

al., 2002). For such experiments, lauroyl chloride was added to a solution of cellulose (0.5 g of cellulose in 25 mL of LiCl/DMAc) and then irradiated in a single-mode microwave reactor for 1 minute with the temperature raised up to 110 °C, which afforded cellulose fatty ester with 1.5 to 2.6 degrees of substitution.

Concerning cellulose fatty ester purification, conventional methods were always based on the use of organic solvents. Application of microwave irradiation resulted in a new method of purification of cellulose fatty ester that relies on their precipitation in sodium hydrogen carbonate after the reaction. The precipitate was filtered and washed with aqueous solution of sodium hydrogen carbonate and then with hot water.

In the next stage, cellulosic plastic films were obtained in homogeneous conditions by microwave-assisted acylation of commercial or chestnut tree sawdust cellulose by fatty acids (Fig. 6.6) (Joly et al., 2005), and the acylation reaction was studied according to the N,N-dimethyl-4-aminopyridine (DMAP) amount, which simultaneously played the role of catalyst and proton-trapping base (Table 6.3).

Obtained long-chain cellulose esters (white or yellow-brown powder from microcrystalline or chestnut tree, respectively) were converted into plastic films by casting from a chloroform solution, which in turn was allowed to evaporate in air at room temperature. Plastic films synthesized in the absence of DMAP showed a decrease in mechanical behavior. Organic (tributylamine) or inorganic bases ($CaCO_3$, Na_2CO_3) were then added to replace DMAP basic activity, and no changes were observed. Thermal and mechanical properties of plastics obtained with various bases, glass transition temperatures (T_g), and degradation temperature (T_d) were found constant whatever the base. The best mechanical properties were obtained for films synthesized in the presence of $CaCO_3$. The same remarks were made concerning the valorization of chestnut tree sawdust cellulose (Table 6.4) (Joly et al., 2005).

The mechanical properties of some largely commercial polymers and cellulose long chain fatty acids are compared in Table 6.4. It can be seen that the cellulose plastic films elastic moduli were always lower than those of

Table 6.3. Cellulose acylation (2 g/100 mL) in the presence of DMAP, after 1 minute of microwave irradiation.

Laurcyl chloride equiv.	DMAP amount					
	0.5 equiv.			0.25 equiv.		
	DS	R_{mass} (%)[a]	R_{mol} (%)[b]	DS	R_{mass} (%)[a]	R_{mol} (%)[b]
2	c	0.8[d]	11	5		
3	1.57	75	27	1.54	80	29
4	1.83	120	39	1.88	178	57
5	2.04	178	54	2.24	270	77
6	2.24	280	79	2.33	290	80
8	2.46	294	78	2.41	293	79
9	2.49	306	80	2.45	322	89
10	2.54	314	81	2.66	330	88

Reprinted from Joly, N., Granet, R., Branland, P., Verneuil, B., Krausz, P. 2005. J. Appl. Polym. Sci. 99:1266, with permission.

[a] Mass yield, ratio between the product weight and starting cellulose weight.

[b] Molar yield, ratio between the mole number of acylated anhydroglucose unit (calculated according to the DS) and mole number of anhydroglucose unit in starting material.

[c] No product was recovered by precipitation.

[d] DS calculated by volumetric method.

traditional plastics; however, their tensile stress and tensile strain level are equivalent to those obtained from linear low-density polyethylene.

Phosphorylation of microcrystalline cellulose under microwave irradiation was also performed (Gospodinowa et al., 2002). The reactions were carried out in a single-mode microwave reactor under argon atmosphere.

The mixtures of 29.0 mmol of urea, 17.6 mmol of phosphorous acid, and 1.8 mmol of cellulose were irradiated for 60 to 120 minutes at the temperature range of 75 to 150 °C (Figure 6.7). The process led to monosubstituted phosphorous acid esters of cellulose of various degrees of substitution of hydroxy functions (0.2 to 2.8) without pretreatment with solvents. The best results (i.e., degree of substitution) were obtained at 105 °C for 2 hours of irradiation (Table 6.5).

As it was presented in Table 6.5, a much higher efficiency and considerable improvements in the degree of substitution of cellulose were seen under microwave irradiation in comparison with conventional heating conditions, which proved the usefulness of microwaves in the modification of natural polymers.

Rapid heating by microwave irradiation was also used for the acid-catalyzed depolymerization of polysaccharides (Fig. 6.8) (Strauss et al., 1995).

Table 6.4.   Mechanical properties of cellulose-based plastic films and some commercial plastic materials.

Base	Elastic modulus E (MPa)	Tensile failure stress σ_f (MPa)	Tensile strain level ε_f (%)
Cellulose-based plastic films with DS = 2.2			
No base	47.4	12.3	314
Tributylamine	54.7	17.0	422
DMAP	46.0	16.4	705
Calcium carbonate	56.3	22.5	687
Some commercial plastic materials			
Cellulose triacetate	1000–4000	12–110	15–55
Cellulose acetate-butyrate	300–2000	20–60	30–60
Low density PE	150	15	500
Low density linear PE	250	20	200–900
High density PE (M ~ 3×10^4 g/mol)	800–1200	35	200–800
High density PE (M = $1 - 5 \times 10^6$ g/mol)	200–600	35	200–500
Polypropylene	1300	40	400

Reprinted from Reprinted from Joly, N., Granet, R., Branland, P., Verneuil, B., Krausz, P. 2005. J. Appl. Polym. Sci. 99:1266, with permission.

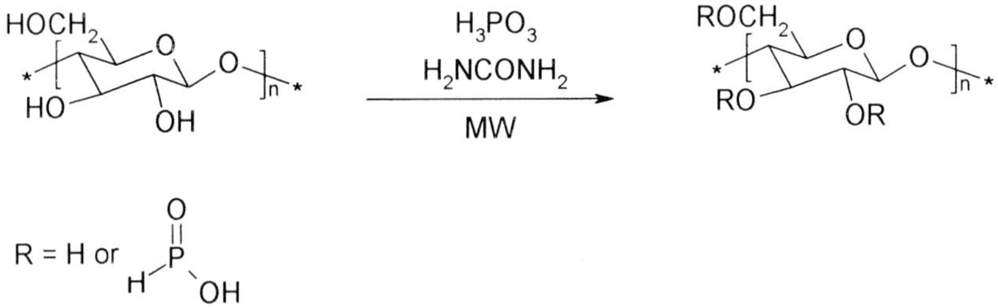

Fig. 6.7.   Phosphorylation of microcrystalline cellulose under microwave irradiation.

High temperatures were required to cleave the glycosidic bonds, but the monosaccharides and oligosaccharides were susceptible to decomposition under the microwave conditions. The reactions were carried out in a pressurized microwave reactor that had a capacity up to 100 mL and was capable of operating at up to 260 °C and 10 MPa (100 bar). The most successful method involved raising the temperature from ambient to 215 °C within

Table 6.5. Data on cellulose phosphorylation.

Type of cellulose	Reaction conditions	Degree of substitution[a]
Cellulose membrane	DMF, phosphorus pentoxide, 25 °C, 48 h	0.6
Hydroxypropyl cellulose	DMF, poly(phosphoric acid) and triethylamine, 120 °C, 6 h	1.3
Poplin cellulose	H_2O, dihydrogen phosphate and urea at 25 °C, post-treatment at 170 °C, 11 min	0.2
Cotton cellulose	DMF, phosphorus oxychloride, 100 °C, 2 h	0.2
Cotton cellulose	Phosphorus acid, 140 °C, 7 h	0.8
Paper cellulose after swelling in a 70% aqueous solution of zinc chloride	Phosphorus acid and urea, 150 °C, 8 h	2.0
Microcrystalline cellulose	Phosphorus acid and urea, MW irradiation, 85 °C, 6 h	0.6
Microcrystalline cellulose	Phosphorus acid and urea, MW irradiation, 105 °C, 2 h	2.8

Reprinted from Gospodinova, N., Grelard, A., Jeannin, M., Chitanu, G.C., Carpov, A., Thiery, V., Besson, T. 2002. Green Chem. 4:220, with permission.

[a] Number of P atoms per cellulose repeating unit.

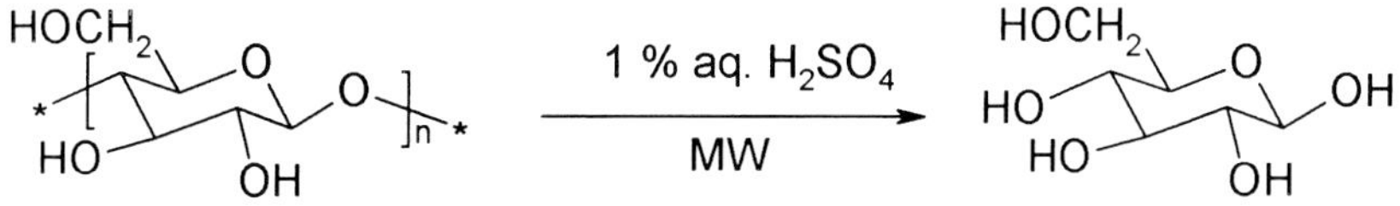

Fig. 6.8. Depolymerization of polysaccharides.

2 minutes, maintaining this briefly, and then cooling. The entire operation was completed with 4 minutes.

As well as cellulose, wood, apart from its traditional uses, can be an attractive source of chemical feedstock. Wood liquefaction is one of the several chemical processes in which wood is derivatized or degraded without costly separation and purification prior to reactions, and as a result, wood is converted into a multifunctional liquid (Burkhardt-Karrenbrock et al., 2001). The process is usually carried out at elevated temperatures, and most

Table 6.6. Efficiency of liquefaction of poplar wood flour in percent conversion under different conditions of microwave irradiation (reaction mixture: 2 g of wood, 10 g of PG, 1 g of MA).

Acid addition (g)	Efficiency of liquefaction (% conversion)					
	500 W				300 W	700 W
	5 min	10 min	15 min	20 min	10 min	10 min
0	—	—	—	—	—	56
0.1	32	41	49	60	—	70
0.2	38	53	68	86	35	86
0.3	40	61	—	99	41	—
0.5	48	82	100	—	42	—
0.7	54	—	—	—	—	—

Reprinted from Krzan, A., Kunaver, M. 2006. J. Appl. Polym. Sci. 101:1051, with permission.

of the investigations have been performed on the use of phenols (Lin et al., 1994, 1997; Maldas et al., 1994) and various polyols (Kobayashi et al., 2004, Krzan et al., 2004). Products of phenol liquefaction were further used in the production of phenol-formaldehyde resins (Lin et al., 2004, 2005) and polyol components in polyurethane foam formulations (Shirashi et al., 1985).

Recently, the liquefaction of wood by glycols (i.e., DEG, EG, and PG) under microwave conditions in the presence of acid anhydrides such as phthalic acid anhydride and trimellitic acid anhydride (TMA) was reported (Krzan et al., 2006). The liquefactions were carried out in a multimode microwave reactor with a temperature sensor (fiberoptic) that was directly inserted into a sealed PTFE reaction vessel (100 mL). A typical reaction mixture that consisted of 2 g of wood, 10 g of glycol, 1 g of acid anhydride, and 0.2 g of mineral acid was irradiated from 5 to 20 minutes with 300 to 700 W of pulse microwave power. After irradiation, the products, which were dark viscous liquids with varying amounts of solid wood remains, were diluted with acetone and filtered in order to separate the liquid from solid particles. The solids were washed and dried to constant weight. The mass of wood remaining after the reaction was taken as the main indication of reaction efficiency. The resulting mass percent conversions are presented in Table 6.6. Acetone was removed from the liquid product in a rotatory evaporator to afford the liquid wood samples with hydroxyl numbers in the range of 500 to 600 mg KOH/g.

These results are favorable compared with the results of wood liquefaction under a similar set of conditions and components with the use of conventional external heating of reaction mixtures that gave conversion in

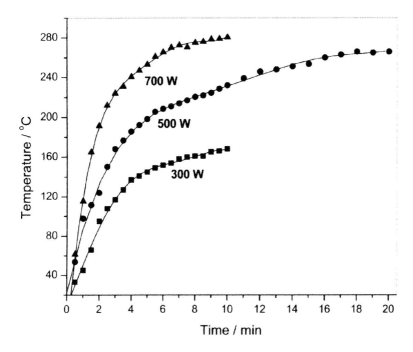

Fig. 6.9. Temperature profiles in reaction vessel under microwave irradiation at different irradiation powers. Reprinted from Krzan, A., Kunaver, M., Tisler, V. 2005. Acta Chim. Slov. 52:253, with permission.

the range of 38% to 56% with reaction times up to 11 hours at 190 °C (Krzan et al., 2005). In the opposite, the presented method allows a rapid heating of the reaction above 250 °C (Fig. 6.9), which mainly contributed to the acceleration of the liquefaction process. For example, when using glycols with organic acid anhydride and phosphoric acid as a catalyst, complete liquefaction was achieved in 20 minutes. However, the obtained liquid wood had a complex composition of low-molecular species, whose chemistry has to be further studied in more detail.

Chitosan

Chitosan, obtained from chitin (i.e., poly-ß-(1→4)-N-acetyl-D-glucosamine) through deacetylation using strong aqueous alkali solutions, is a more versatile form of this polysaccharide, which is one of the most abundant natural polymers on Earth after cellulose. Much attention has been paid to

Table 6.7.   Molecular weight after microwave and conventional heating.

Entry	Ionic strength (I)	Molecular weight ($\times 10^4$)		
		NaCl	KCl	CaCl$_2$
1	0	10.5		
2	0.01	3.52	2.79	3.09
3	0.01[a]	51.5	50.7	51.0

Reprinted from Xing, R., Liu, S., Zu, H., Guo, Y.,Wang, P., Li, C., Li, Y., Li, P. 2005. Carbohydr. Res. 340:2150, with permission.

[a] With conventional heating to replace microwave irradiation to maintain the reaction volume for 2 h.

modification and utilization of chitosan due to its good biodegrability, bio-compatibility, and bioactivity (Kumar, 2000; Kurita, 2001; Roberts, 1992).

Recently, the hydrolysis of chitosan from shrimp shells (degree of deacetylation of 0.85) into oligosaccharides under a microwave irradiation was investigated in the presence of inorganic salts such as sodium chloride, potassium chloride, and calcium chloride (Xing et al., 2005). The reactions were carried out in a domestic microwave oven without any modification, in which a 300-mL Erlenmeyer flask containing 100 mL of a chitosan solution in HCl (2% w/v) or acetic acid (2% w/v) with an inorganic salt (0.15 mol/L Cl$^-$) was irradiated up to approximately 100 °C for 0.5 to 25 minutes with 480 to 800 W of microwave power. It was found that the molecular weight of the degraded chitosan obtained by microwave irradiation was considerably lower than those obtained by conventional heating (2 hours) (Table 6.7).

The results show that the degradation of chitosan under both microwave and conventional conditions was influenced by the addition of inorganic salts. For example, under microwave irradiation, after 25 minutes, the molecular weights of degraded chitosan were approximately 10×10^4 g/mol and 3×10^4 g/mol in the absence and presence of inorganic salts, respectively. Microwave irradiation significantly accelerated the degradation of chitosan in comparison to conventional heating that resulted in molecular weight of degraded chitosan approximately 51×10^4 g/mol. Moreover, microwave-assisted degradation in the presence of inorganic salts did not affect the pyranose rings of chitosan oligosaccharides compared to products of traditional technology (Xing et al., 2005).

More recently, the graft polymerization of ε-caprolactone onto chitosan (degree of deacetylation of 1.0) was carried out under microwave irradiation (Liu et al., 2005). The copolymerization reactions were run via a protection-

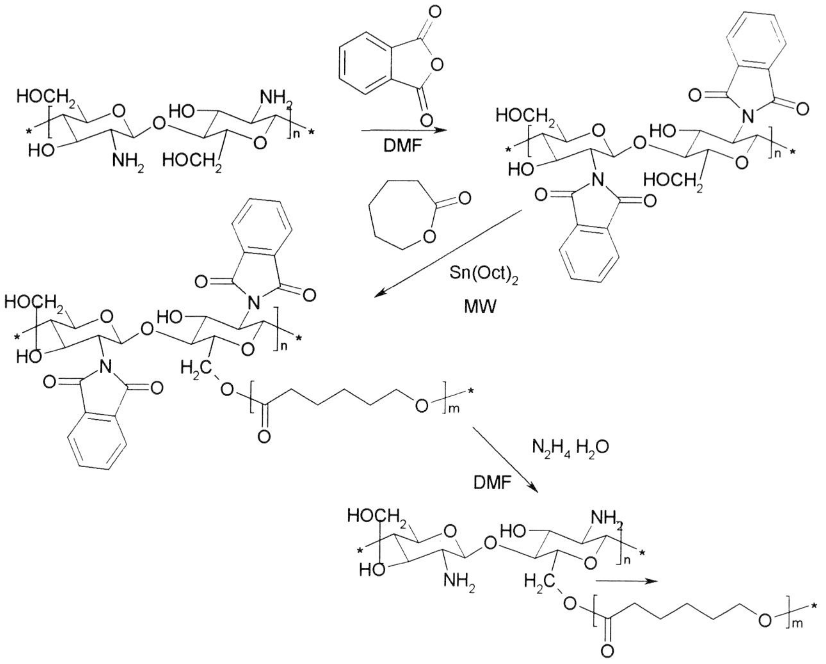

Fig. 6.10. The copolymerization reactions via a protection-graft-deprotection procedure with phthaloylchitosan (PHCS) as a precursor and stannous octoate as a catalyst.

graft-deprotection procedure with phthaloylchitosan (PHCS) as a precursor and stannous octoate as a catalyst (Fig. 6.10).

The reactions were carried out in a household microwave oven, in which PHCS (1 g) was placed in a dried glass reactor while a certain amount of the mixture of ε-caprolactone monomer and stannous octoate (1 mol% to ε-caprolactone) was poured into the reactor. The reactor was sealed under high-purity (99.999%) nitrogen, and preswelling of PHCS with monomer was conducted overnight. Then the reactor was irradiated for 15 minutes with 450 W of microwave power. Next, the phthaloyl-protected graft copolymer (1 g) was stirred in 20 mL of water and heated up to 100 °C for 2 hours under nitrogen while hydrazine monohydrate was added to deprotect the phthaloyl groups. The solution was allowed to cool to room temperature in precipitate of the copolymer. After deprotection, the phthaloyl groups were removed so the amino groups were regenerated, and chitosan-g-polycaprolactone with high grafting percentage even above 100% was achieved (Table 6.8). Thus, the chitosan-g-polycaprolactone copolymer

Table 6.8. Synthesis of chitosan-g-polycaprolactone under microwave irradiation.

| | | Chitosan-g-polycaprolactone | |
Trial[a]	ε-Caprolactone (g)	Yield (g)	Grafting percentage[b] (wt%)
1	0.6	0.6	10.0
2	1.0	0.7	25.8
3	1.6	0.9	41.8
4	2.0	1.0	120
5	3.0	1.1	232

Reprinted from Liu, L., Li, Y., Fang, Z., Chen, L. 2005. Carbohydr. Polym. 60:351, with permission.

[a] Feed amount of PHCS, 1 g. Microwave irradiation, 450 W for 15 min.

[b] Grafting percentage of polycaprolactone = [(weight of introduced polycaprolactone branches)/(weight of chitosan main chain)] × 100, determined by 1H NMR.

was an amphoteric hybrid with both a large amount of free amino groups and hydrophobic polycaprolactone side chains.

The results of copolymerization with varying amounts of ε-caprolactone show that the yield and grafting percentage of polycarprolactone increased with the ε-caprolactone amount in the reaction mixture. For instance, the graft copolymers having rather high grafting percentages of 120% and 230% were obtained when they were irradiated at a microwave oven with 450 W of microwave power. When the power output was above 450 W, the graft copolymerization was greatly improved but discoloration of the copolymers was observed. In comparison, under conventional conditions, similar results can be obtained after 20 hours or more of heating (Kurita et al., 2002; Yoksan et al., 2001).

In turn, it was shown that polyacrylonitrile can be grafted onto chitosan under microwave irradiation (Singh et al., 2005). For this purpose, a solution of chitosan (0.1 g) in 25 mL of 5% aqueous solution of formic acid (v/v) and acrylonitrile (26×10^{-2} mol/L) was irradiated in a household microwave oven in a 150-mL flask when the average bulk temperature at the end of the reaction was measured by inserting a thermometer into the reaction mixture. The reaction was repeated with different monomer concentrations in the range of 10 to 28×10^{-2} mol/L. Finally, applying microwave irradiation, polyacrylonitrile was grafted onto chitosan with 170% grafting yield under homogeneous conditions in 1.5 minutes in the absence of any radical initiator or catalyst. It was found that under similar conditions, a maximum grafting of 105% could be achieved when the $K_2S_2O_8$/ascorbic

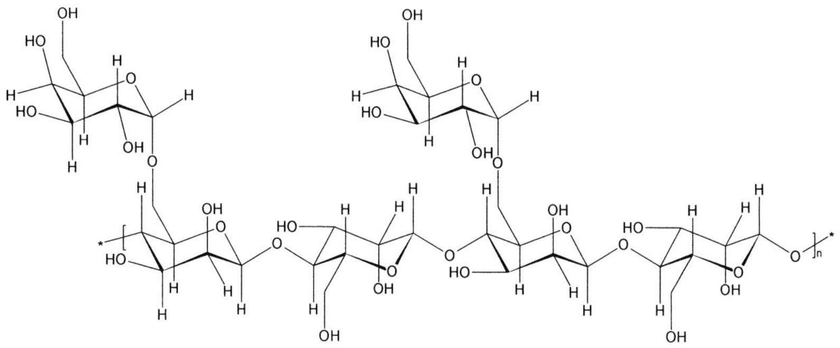

Fig. 6.11. Branched chain polymer consisting of straight chain of mannose units joined by ß-D-(1→4) linkages having α-D-galactopyranose units attached to this linear chain by (1→6) linkages.

acid redox system was used as radical initiator in a thermostatic water bath at 35 °C. The grafting was found to increase with an increase in the initial concentration of monomer in the range of 10 to 28 10^{-2} mol/L. Also, grafting was found to increase up to 80% microwave power and thereafter decrease, which may be due to either more homopolymerization or some decomposition of grafted copolymer taking place at microwave power higher than 80%.

Guar Gum

In a similar way, under microwave irradiation, polyacrylamide was grafted onto guar gum (Singh et al., 2004), which is edible carbohydrate polymer extracted from the seed of guar bean *(Cyanaposis tertagonolbus)*. It is a branched chain polymer consisting of straight chain of mannose units joined by ß-D (1→4) linkages having α-D-galactopyranose units attached to this linear chain by (1→6) linkages (Fig. 6.11).

Guar gum forms viscous colloidal dispersion when hydrated in cold water and is used as a thickener and binder of free water in industrial and food applications; however, its viscosity is difficult to control because of biodegradation, which limits applications (Deshmukh et al., 1985, 1987). Therefore, modification of guar gum by grafting of water-soluble monomers results in the retention of desirable properties and incorporation of favorable properties (Sharma et al., 1999).

Under microwave conditions, a solution (25 mL) of guar gum (0.1 g), acrylamide (16 × 10^{-2} mol/L), AgNO$_3$ (8.0 × 10^{-5} mol/L), K$_2$S$_2$O$_8$ (1.0 × 10^{-3} mol/L), and ascorbic acid (22 × 10^{-2} mol/L) in water in a 150-mL

Table 6.9.   Optimum conditions to get maximum %G.

	Conventional method with redox system and catalyst	Under microwave, presence of redox system and catalyst	Under microwave, absence of redox system and catalyst
% Grafting	140	190	120
% Efficiency	49.1	66.7	42.1
Microwave power, %	—	80	70
Temperature, °C	60	60	63
Time, min	80	0.22	0.33
% N	4.98	3.26	2.47

Reprinted from Singh, V., Tripathi, D.N., Tiwari, A., Sanghi, R. 2004. Carbohydr. Polym. 58:1, with permission.

flask was irradiated in a domestic microwave oven. The average bulk temperature was measured at the end of the reaction by inserting a thermometer in the reaction mixture. Because all the experiments were performed with water as a solvent, temperatures were less than 100 °C. For example, after 0.22 minute and 0.33 minute of irradiation with 960 W of microwave power, the temperature reached 60 °C and 73 °C, respectively.

Guar-g-polyacrylamide (G-g-PAA) was separated by precipitating the reaction mixture with methanol/water. Graft copolymerization of the guar gum with acrylamide under the influence of microwave irradiation was also carried out in the absence of any radical initiators and catalyst (i.e., $K_2S_2O_8$/ascorbic acid) and resulted in grafting yields comparable to redox ($K_2S_2O_8$/ascorbic acid) initiated by conventional heating (Table 6.9) (Singh et al., 2004).

The grafting efficiency up to 20% was further increased when initiators and catalysts were used with microwave irradiation. Maximum grafting efficiency achieved under microwave method was 66% in 0.22 minute in comparison to 49% in 90 minutes by the conventional method.

Most recently, in the same manner, acrylonitrile applied without any redox/radical initiator was grafted onto seed gum from *Cassia siamea* (Singh et al., 2006), which is a complex polysaccharide containing D-glucose, D-mannose, and D-xylose as components in a 3:6:2 ratio (Khare et al., 1980). The influence of microwave power and exposure time on the graft copolymerization was studied and the maximum percent grafting (%G) and efficiency (%E) were found 150% and 43.5%, respectively, after 2 minutes of irradiation (Table 6.10).

Table 6.10. %G and %E with change in MW power and exposure time at fixed concentration of acrylonitrile (26×10^{-zM}), CS gum (4 g/L), and total reaction volume (25 mL).

MW power	Temperature (°C)	Exposure Time (s)	%G	%E
	52	30	30	8.7
	62	60	40	11.61
	81	90	45	13.06
40%	86	120	55	15.96
	55	30	40	11.61
	65	60	52	15.09
	84	90	68	19.73
60%	88	120	90	26.12
	62	30	55	15.96
	70	60	95	27.57
	90	90	105	30.47
80%	95	120	120	34.83
	80	30	65	18.86
	92	60	92	26.70
	95	90	115	33.38
100%	95	120	150	43.54

Reprinted from Singh, V., Tripathi, D.N. 2006. J. Appl. Polym. Sci. 101:2384, with permission.

At the same monomer concentration, in the presence of redox ($K_2S_2O_8$/ascorbic acid) initiator, grafting of acrylnitrile onto the seed gum resulted in 80% of grafting with 23.3% efficiency under conventional conditions. Interestingly, when a radical scavenger (i.e., hydroquinone) was added to the reaction mixture for the microwave as well as conventional experiments, no grafting of acrylnitrile onto the seed gum was observed, which proved there was the same radical mechanism of the reactions under both microwave and conventional conditions.

Starch

Among the industrial materials, starch is one of the cheapest and most cost-effective substrate. It is used as coating and sizing in paper, textiles, and carpets, as binders and adhesive, and as absorbents and encapsulants (Whistler et al., 1984, 1993).

Effective starch modification can be achieved when microwave irradiation corroborates with the activity of proton originating from mineral acid added to the reaction mixture. For example, such an application has been recommended for fast and efficient saccharification of sugar syrups of high dextrose equivalent (Azuma et al., 1989). Fully amorphous dextrin may result from microwave irradiation of maize starch of 4% to 60% moisture content to 90 °C for 40 seconds (Marquette et al., 1982).

Maize, cassava, and potato starch of 10.3%, 10.2%, and 14.5% humidity, respectively, were also modified under microwave irradiation with $TiCl_4$, formaldehyde, and acetylene, while plain starch underwent dextrinization (Muzimbaranda et al., 1994). Starch was irradiated in a household microwave oven in the air-dried state as well as 4% to 7% slurries in distilled water. Irradiation of starch was also carried out in either 1% to 10% (w/w) of 40% aqueous formaldehyde or 1% to 10% of $TiCl_4$. Moreover, starch was irradiated in the atmosphere of acetylene with or without addition of either $TiCl_4$ (5%, w/w) or hydrogen peroxide (6%, w/w), while the temperature of the reaction vessel was in the range of 80 to 90 °C.

The decomposition of maize and cassava starch, which have a similar water content (i.e., 10.2% and 10.3%, respectively), began almost immediately after setting starch on microwave irradiation, whereas the first 5 minutes of irradiation of potato starch brought no changes in its granules expressed by viscosity changes (Fig. 6.12). After the induction period, potato starch underwent dramatic changes, and the viscosity of its gel decreased by one order of magnitude, which was explained not only of the basis of water content but also by taking into account the dimension of granules, water distribution in their interior, and amylose-to-amylopectin ratio.

Upon microwave irradiation, starch blended with $TiCl_4$ and gave gels of low viscosity, while reaction with formaldehyde as well as with acetylene gave products of enhanced viscosity (Muzimbaranda et al., 1994).

Recently, normal maize, waxy maize, and amylomaize V starches were treated at a moisture content of 30% under microwave irradiation (Luo et al., 2006). The moisture content was adjusted to 30% by the addition of an appropriate amount of water to the starch samples of known moisture content. Then the moistened starch samples were placed in glass beakers and sealed with a perforated polyethylene film designed for microwave ovens. Irradiation was carried out for 20 minutes with a household microwave oven.

The results showed that microwave treatment induced the rearrangement of molecules in sections of starch that are part of the crystalline region and that irradiation had a more pronounced effect on amylose than on amylopectin. DSC data showed that the imperfect crystallites melted and recrystallized into more perfect ones. The stability of these newly formed crystallites was different from those present in the native starches. The extent of the changes induced by microwave treatment depended not only on the crystal structure of starch but also on its amylose content (Luo et al., 2006).

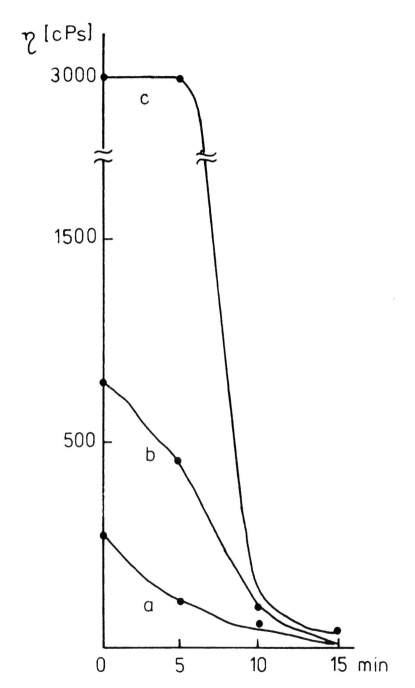

Fig. 6.12. The viscosity decrease of 5% gels made or air-dried maize (**a**), cassava (**b**), and potato (**c**) starch irradiated in the period from 0 to 15 minutes. Reprinted from Muzimbaranda, C., Tomasik, P. 1994. Starch 46:469, with permission.

Table 6.11.   Effect of temperature on composition and properties of starch acetates made from dry corn starch/glacial acetic acid/acetic anhydride 1:1:0.5 (w/w/w).

Temperature program[a]		DS[b]	RE (%)[c]	WSI (%)[d]	WAI (g/g)[e]
Ramp 1 (T, t)	Ramp 2 (T, t)				
25—120, 3.5	120, 10	0.19	24	12	10.6
25—120, 3.5	120, 20	0.21	26	15	13.1
25—120, 3.5	120, 40	0.30	38	2	13.6
25—150, 3.5		0.33	42	49	14.4
25—150, 3.5	150-160, 1.5	0.57	72	76	11.0
25—150, 3.5	150-160, 2.5	0.76 ± 0.02[f]	96 ± 3	91 ± 4	2.4 ± 0.9
25—150, 3.5	150-160, 5.0	0.74	93	93	1.4
25—175, 3.5		0.77	97	99	0.2

[a] T = temperature (°C); t = time (min).
[b] Degree of substitution.
[c] Reaction efficiency (based on acetic anhydride)
[d] Water solubility index.
[e] Water absorption index.
[f] Error estimates shown represent the mean and standard deviation from three repetitions of the experiment.
Reprinted from Shogren, R.L., Biswas, A. 2006. Carbohydr. Polym. 64:16, with permission.

Starch acetates of degree of substitution (DS) of 0.1 to 1.5 were prepared by heating corn starch, acetic acid, and acetic anhydride in a sealed vessel in a microwave reaction (Shogren et al., 2006). Typically, 70 g of starch, 70 g of acetic acid, and 0 to 70 g of acetic anhydride are added to a 270-mL Teflon vessel. A magnetic stir bar was added, and the mixture was stirred for 5 minutes. The vessel was then sealed, the thermocouple was inserted, and the vessel was irradiated in a multimode microwave reactor. The temperature ramp was 25 to 150 °C over 3.5 minutes to 150 to 160 °C over 1.5 to 4.5 minutes. Maximum microwave power was limited to 650 W to avoid overshooting the temperature program. After opening the reactor, the contents were placed in a Waring blendor with 500 mL of ethanol and blended until the precipitate was broken into fine particles. The effect of temperature on extent of acylation and solubility/swelling properties was studied (Table 6.11).

The reaction efficiencies (RE) were typically over 90% at the reaction temperatures of 150 to 160 °C for 4 to 7 minutes. Starch acetates prepared by reaction at 160 °C to DS values of 0.4 to 0.9 were soluble in room-temperature water in more than 90%. Colors of the starch acetates prepared

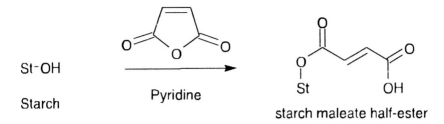

Fig. 6.13. Reaction of starch and maleic anhydride in the presence of pyridine as a catalyst.

by this method were white to light tan. Heating for longer times at 160 to 175 °C gave products that became more brown in color (Shogren et al., 2006a).

In turn, sodium acrylate was grafted onto corn starch under microwave irradiation to obtain a superabsorbent (Tong et al., 2005). Potassium persulfate (PPS) and polyethyleneglycol diacrylate were used as an initiator and crosslinker, respectively. In a 100-mL PTFE beaker, starch was mixed with the crosslinker solution (0.4 vol%) and initiator (0.1 mol/L). Sodium acrylate was prepared by dropping NaOH slowly to acrylic acid while stirring in a cool water bath. Then, sodium acrylate was added to the beaker and thoroughly mixed until the temperature of the mixture was approximately 30 °C. The beaker was placed in a domestic microwave oven and irradiated for 8 to 12 minutes with 85 to 115 W of microwave power. It was found that microwave irradiation could substantially accelerate the synthesis without needing to remove O_2 or inhibitor. Moreover, microwave power was believed to be the most significant factor affecting the swelling ratio and solubility of the product. Optimized experimental results showed that 10 minutes of microwave irradiation at 85 to 90 W could produce a corn starch–based superabsorbent with a swelling ratio of 520 to 620 g/g in distilled water and solubility of 8.5 to 9.5 wt%.

More recently, the method for the preparation of starch maleate esters in the reaction of starch and maleic anhydride in the presence of pyridine as a catalyst under microwave irradiation was described (Fig. 6.13) (Biswas et al., 2006).

The reactions were performed in a 50-mL quartz vessel, in which a mixture of starch contained 11% moisture (3 g, approximately 17 mmol of anhydroglucose), 0.91 g [8.5 mmol] of maleic anhydride, 0.75 mL [8.5 mmol] of pyridine, and 6 mL of DMSO were stirred to form a smooth suspension. The mixture was irradiated in a multimode microwave reactor so that temperature reached 100 °C in 2.5 minutes and was maintained for another 2.5 minutes. The starch maleate esters were precipitated from isopropyl

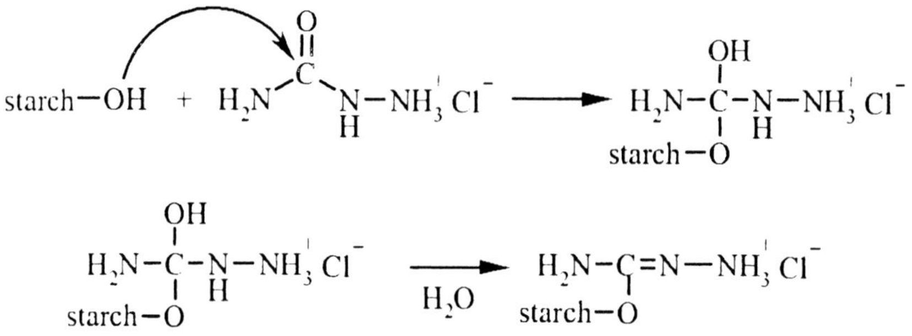

Fig. 6.14.   Reaction of potato starch with semicarbazide hydrochloride under microwave irradiation.

alcohol to obtained starch with DS up to 0.25 with a reaction efficiency of 50%. In absence of pyridine, DS reached 0.1, but when maleic anhydride was substituted with succinic anhydride or octenyl succinic anhydride, DS increased to 0.3. Under conventional conditions with the same heating rate and time, the starch maleate was afforded with DS 0.13 in 80% yield compared to DS 0.25 and 90% yield obtained from the microwave experiments.

O-(C-carbamoylated) starch was obtained when potato starch (11% moisture) was irradiated together with semicarbazide hydrochloride (Fig. 6.14) (Siemion et al., 2005).

The reagents (i.e., starch and semicarbazide hydrochloride) were blended in 1:1 (mol/mol of D-glucose unit) proportion and either heated in the air for 5 minutes in a domestic microwave oven at 900 W or in a furnace at 205 °C for 1 hour. It was found that the products retained their hydrochloride salt character, making them cationic (Table 6.12).

However, the reaction time under microwave irradiation was much shorter in comparison to conventional conditions (i.e., 5 minutes versus 1 hour); both methods provided close to the same cationic products. The conventional heating damaged starch granules to a higher extent than microwave irradiation, while the degree of the reaction was slightly higher under microwave conditions.

The hydrolysis of starch to glucose under microwave irradiation was also examined (Yu et al., 1996). For such a purpose, starch (10%) was suspended in 1 mL of diluted hydrochloric acid (0.5 M) and irradiated for 6 minutes in a domestic microwave oven with 130 W to 400 W of microwave power so that temperature reached 95 °C. For reactions under conventional conditions, the starch suspension (4.0 mL) was placed in a sealed tube, purged with nitrogen, and heated at 100 °C for 60 minutes. The reaction mixtures were analyzed by means of HPLC, applying samples of dextran

Table 6.12. Solubility in water, pH, and conductivity of 1% aqueous solution of *O*-(C-carbamoylated) starches.

Sample	pH	Conductivity (µS/cm)	Solubility (%) (°C) 25	50	75
Original starch	7.03	13		2.0	
Conventionally heated starch[a]	7.12	30	1.4	7.4	72.4
Microwave-heated starch	6.22	123	89.0	92.0	93.4
Semicarbazide hydrochloride	2.53	9980			
Product from conventional heating	2.95	3790	9.4	9.7	13.1
Product from microwave heating	2.96	4860	8.4	9.8	12.1

[a] Potato starch processed identically but without semicarbazide hydrochloride. Reprinted from Siemion, P., Kapusniak, J., Koziol, J. 2005. Carbohydr. Polym. 62:182, with permission.

(M_W 40,000 g/mol), dextran (M_W 8800 g/mol), and glucose as references (Fig. 6.15).

After microwave irradiation, the reaction mixture was clear and no retrograded starch remained in a suspension. In the opposite, following 60 minutes of conventional heating, the soluble starch was completely converted to glucose and the retrograded starch remained suspended in the solution, which led to the conclusion that microwave irradiation caused deformation of ß-structure of the retrograded starch, making it soluble in the solution and, therefore, subject to hydrolysis. To prove the conclusion, the isolated retrograded starch from the conventional heating experiment was resuspended in either water or dilute hydrochloric acid and irradiated for 5 minutes. The retrograded starch became soluble in water and was completely hydrolyzed in hydrochloric acid. Moreover, under conventional conditions, the reaction solution became colored and its absorbance increased significantly at wavelength between 400 and 500 nm, while under microwave irradiation the absorbance of the starch solution was near zero.

Recently, microwave irradiation was applied for chemical derivatizations of polysaccharides (i.e., potato starch, chitosan, and Konjac) to obtain a medium to high degree of substitution for starch acetates, starch succinates, carboxymethyled Konjac, aminated starch, and aminated chitosan. Then, the microwave-assisted ring-opening polymerization of cyclic monomers (i.e., ε-caprolactone) was successfully extended to grafting reactions between different biodegradable polymers, such as starch and polycaprolactone (Table 6.13) (Koroskenyi et al., 2002).

Acetylation was carried out by mixing starch with acetic anhydride and catalyst in a beaker and irradiating it in a household microwave oven (600

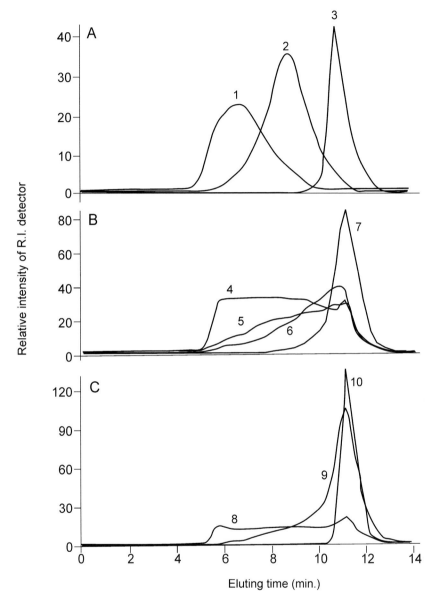

Fig. 6.15. Analysis of hydrolysate of starch solution after microwave irradiation or thermohydrolysis in acidic suspension. HPLC profiles of (1) dextran (M_w = 40,000); (2) dextran (M_w = 8800); (3) glucose; (4–7) starch suspension incubated at 100 °C for 5, 10, 20, and 60 minutes, respectively; and (8–10) starch suspension subjected to microwave irradiation at 40% power for 3, 4, 5 and minutes, respectively. Reprinted from Yu, H.M., Chen, S.T., Suree, P., Nuansri, R., Wang, K.T. 1996. J. Org. Chem. 61:9608, with permission.

Table 6.13. Reaction conditions and results for the synthesis of caprolactone graft polymers.

#	Starting Material	Preswelling	Grafting Efficiency (%)	Degree of grafting (%)
GCL1	3.0 g Potato starch	CL overnight	0	0
GCL2	3.0 g Potato starch	7 mL water overnight	24	25
GCL3	3.0 g Potato starch	7 mL water+CL overnight	21	22
GCL4	1.0 g Konjac acctate	—	7.1	22

Reprinted from Koroskenyi, B., McCarthy, S.P. 2002. J. Polym. Environm. 10:93, with permission.

Table 6.14. Carboxymethylation of Konjac with chloroacetic acid.

#	Konjac (g)	KOH (g)	$ClCH_2COOH$ (g)	Solvent (mL)	Time (s)
CA1	1.0	1.0	1.0	10.0	35
CA2	1.0	1.0	1.0	15.0	50
CA3	1.0	1.0	1.0	10.0	50
CA4	1.0	1.0	1.0	15.0	70
CA5	5.0	5.0	5.0	40.0	90

Reprinted from Koroskenyi, B., McCarthy, S.P. 2002. J. Polym. Environm. 10:93, with permission.

W or 1300 W). The product was thoroughly washed with acetone/water (1:1) twice and then once with acetone and dried in vacuum.

Konjac was carboxymethylated with chloroacetic acid by dissolving potassium hydroxide and chloroacetic acid in 80% aqueous isopropanol, adding Konjac, and following the reaction at room temperature for 24 hours; it was then irradiated after thorough mixing. For comparison, conventional heating was carried out as follows: 30.0 g of Konjac was suspended in 450 mL of 80% aqueous isopropanol to which 36 g of sodium hydroxide and 30 g of chloroacetic acid were added. The mixture was stirred at 80 °C for 1 hour. After the mixture cooled to room temperature, it was neutralized with hydrochloric acid and then precipitated with ethanol/acetone (1:2) and dried in vacuum. The product was completely soluble in water in contrast to native Konjac (Table 6.14). Carboxymethylation with glyoxylic acid was carried out according to the following pro-

Table 6.15.   Reaction conditions for the amination of starch with CHPTMAC.

#	CHPTMAC (g)	pH	Water (mL)	Power (W)	Time (min)	DS
GMACS1	7.8	(12% NaOH)	18	260	3	1.54
GMACS2	15.6	12	106	390	10	0.83
GMACS3	15.6	12	106	650	3	0.64
GMACS4	15.6	12	106	1300	1	0.21

Reprinted from Koroskenyi, B., McCarthy, S.P. 2002. J. Polym. Environm. 10:93, with permission.

Table 6.16.   Reaction conditions and results for the amination of starch with DEAEC.

#	NaOH (g)	Water (mL)	Power Time (W)	(min)	DS
DEAEC S1	(pH = 12)	~100	390	10	1.01
DEAEC S2	5.0	10	260	3	0.34
DEAEC S3	2.5	5	260	3	1.24

Reprinted from Koroskenyi, B., McCarthy, S.P. 2002. J. Polym. Environm. 10:93, with permission.

cedure: 1.0 g of Konjac was placed in a 20-mL vial, 5.0 mL of a 50% gly-oxylic acid solution was added, and the mixture was irradiated for 2 minutes. The product was washed with water, precipitated with methanol, and dried in vacuum. It was completely dissolved in water (in contrast to native Konjac) and precipitated in the form of fine powder (Koroskenyi et al., 2002).

A general procedure for the amination with (3-chloro-2-hydroxypropyl)-trimethylammonium chloride (CHPTMAC) was 80 mL of NaOH solution (pH 12) charged into a 400-mL beaker to which 26.0 mL of 60% solution of CHPTMAC was added. The pH was adjusted with 3 M NaOH solution until it remained constant (12.0); 2.0 g of starch or chitosan was added, and the mixture was irradiated in a microwave oven. The product was acidified with hydrochloric acid, dialyzed for several days, and finally freeze-dried (Table 6.15) (Koroskenyi et al., 2002).

Amination with 2-(diethylamino)ethyl chloride hydrochloride (DEAEC) was carried out as follows: 8.0 g of DEAEC was added to 80 mL of NaOH solution at pH 12.0. The pH was adjusted to 12.0 with 3 M NaOH solution until it remained constant; 2.0 g of starch or chitosan was added, and the mixture was irradiated in a microwave oven, with pressure built up in a closed vessel. The product was acidified with hydrochloric acid, dialyzed for several days, and finally freeze-dried (Table 6.16) (Koroskenyi et al., 2002).

References

Azuma, J., Tomiya, T., Katada, T. 1989. Jpn. Kokai Tokkyo Koho, Jp. 01 225 601.

Badri, K.H., Ahmad, S.H., Zakaria, S. 2001. J. Appl. Polym. Sci. 81:384.

Biswas, A., Shogren, R.L., Kim, S., Willet, J.L. 2006. Carbohydr. Polym. 64:484.

Bogdal, D., Prociak, A., Kruszec, A. 2006. unpublished data.

Burkhardt-Karrenbrock, A., Seegmulleer, S., Burk, R. 2001. Holz. Roh. Werkstoff 59:13.

Chian, K.S., Gan, L.H. 1998. J. Appl. Polym. Sci. 68:509.

Deshmukh, S.R., Chaturvedi, P.N., Singh, R.P. 1885. J. Appl. Polym. Sci. 30:4013.

Deshmukh, S.R., Singh, R.P. 1887. J. Appl. Polym. Sci. 33:1963.

Gospodinova, N., Grelard, A., Jeannin, M., Chitanu, G.C., Carpov, A., Thiery, V., Besson, T. 2002. Green Chem. 4:220.

Gourson, C., Benhaddou, R., Granet, R., Krausz, P., Saulnier, L., Thibault, J.F. 1999. C. R. Acad. Sci. Paris 2:75.

Guo, A., Jani, J., Petrovic, Z. 2000. J. Appl. Polym. Sci. 77:467.

Heinze, T.J., Glasser, W.G. 1998. ACS Symposium Series. Chapter 3. 688:38.

Hofer, R. 1999. Oleochemical polyols: New raw materials for polyurethane applications. Proceedings European Coatings Conference, Berlin.

Hong, H.Y., Gao. Y., Wang, D.N., Hu, C.P., Zu, S., Vanoverloop, L., Randall, D. 2002. J. Appl. Polym. Sci. 84:591.

Huy, H.T., Buu, T.N., Dung, T.T.K., Han, T.N., Qaung, P.V. 1996. J. Mater. Sci., Pure Appl. Chem. A33:1957.

Joly, N., Granet, R., Branland, P., Verneuil, B., Krausz, P. 2005. J. Appl. Polym. Sci. 99:1266.

Khare, N., Dubey, P., Gupta, P.C. 1980. Planta Med. 40(suppl.):76.

Kobayashi, M., Asano, T., Kajiyama, M., Tomita, B. 2004. J. Wood Sci. 50:407.

Koroskenyi, B., McCarthy, S.P. 2002. J. Polym. Environm. 10:93.

Krzan, A., Kunaver, M. 2006. J. Appl. Polym. Sci. 101:1051.

Krzan, A., Kunaver, M., Tisler, V. 2005. Acta Chim. Slov. 52:253.

Kumar, M.N.V.R. 2000. React. Funct. Polym. 46:1.

Kurita, K. 2001. Prog. Polym. Sci. 26:1921.

Kurita, K. Ikeda, H., Yoshida, Y., Shimojoh, M., Harata, M. 2002. Biomacromolecules 3:1.

Lin, L., Yao, Y., Yoshioka, M., Shiraishi, N. 1997. J. Appl. Polym. Sci. 64:351.

Lin, L., Yoshioka, M., Yao, Y., Shiraishi, N. 1994. J. Appl. Polym. Sci. 52:1629.

Lin, L., Yoshioka, M., Yao, Y., Shiraishi, N. 1995. J. Appl. Polym. Sci. 58:1297.

Liu, L., Li, Y., Fang, Z., Chen, L. 2005. Carbohydr. Polym. 60:351.

Luo, Z., He, X., Fu, X., Luo, F., Gao, Q. 2006. Starch 58:468.

Maldas, D., Shirashi, N. 1997. Biomass Bioenergy 12:273.

Marquette, G.H.A., Gonze, M., Lane, C. 1982. Eur. Pat. Appl. EP 59 050.

Mazzocchia, C., Kaddouri, A., Modico, G., Nannicini, R., Martini, F., Marengo, S. 2005. Fatty Acid Methyl Esters Synthesis from Triglycerides over Homogenous/Heterogeneous Catalysts in the Presence of Microwaves. Proceedings of 10th International Conference on Microwave and RF Heating. Modena, Italy.

Mikolajek, B. 2006. Master Thesis, Politechnika Krakowska (Cracow University of Technology), Krakow, Poland.

Muzimbaranda, C., Tomasik, P. 1994. Starch 46:469.

Prociak, A. 2006. Rigid polyurethane foams modified with vegetable oil-based polyols. UTECH 2006 Conference Papers on CD, Maastricht, The Netherlands.

Roberts, G.A.F. 1992. Chitin chemistry. MacMillan, London.

Satge, C., Verneuil, B., Branland, P., Granet, R., Krausz, P., Rozier, J., Petit, C. 2002. Carbohydr. Polym. 49:373.

Sharma, B.J., Kumar, V., Soni, P.L. 1999. Trends Carbohydr. Chem. 5:5.

Shirashi, N., Onodera, S., Ohtani, M., Masumoto, T. 1985. Mokuzai Gakkaishi 31:418.

Shogren, R.L., Biswas, A. 2006. Carbohydr. Polym. 64:16.

Siemion, P., Kapusniak, J., Koziol, J. 2005. Carbohydr. Polym. 62:182.

Singh, V., Tripathi, D.N. 2006. J. Appl. Polym. Sci. 101:2384.

Singh, V., Tripathi, D.N., Tiwari, A., Sanghi, R. 2004. Carbohydr. Polym. 58:1.

Singh, V., Tripathi, D.N., Tiwari, A., Sanghi, R. 2005. J. Appl. Polym. Sci. 95:820.

Strauss, C.R., Trainor, R.W. 1995. Aust. J. Chem. 48:1665.

Tong, Z., Peng, W., Zhiqian, Z. Baoxiu, Z. 2005. J. Appl. Polym. Sci. 95:264.

Whistler, R.L., BeMiller, J.N. 1993. Industrial polysaccharides and their uses. San Diego, Academic Press.

Whistler, R.L., BeMiller, J.N., Paschall, E.F. 1984. Starch: Chemistry and technology. Orlando, Academic Press.

Wiesbrock, F., Hoogenboom, R., Abeln, C.H., Schubert, U.S. 2004. Macromol. Rapid Commun. 25:1895.

Xing, R., Liu, S., Zu, H., Guo, Y.,Wang, P., Li, C., Li, Y., Li, P. 2005. Carbohydr. Res. 340:2150.

Yoksan, R., Akashi, M., Biramontri, S., Chirachaanchai, S. 2001. Biomacromolecules 2:1038.

Yu, H.M., Chen, S.T., Suree, P., Nuansri, R., Wang, K.T. 1996. J. Org. Chem. 61:9608.

Zlatanic, A., Lava, C., Zhang, W., Petrovic, Z. 2004. J. Polym. Sci., Part B: Polym. Physics, 42:809.

7

RECYCLING OF PLASTICS

Polymer degradation and decomposition is a subject of current interest to industry and researchers because of environmental reasons including the increase of plastic waste. Nowadays, the investigation for new energy-efficient solutions for recycling processes of different plastic materials results in the application of microwave irradiation and other unconventional techniques for performing chemical reaction by chemical engineers. The recycling processes demand the material be heated and therefore has led to the use of microwaves, which offer increased reaction rates and energy savings in comparison with thermal heating.

In the United Kingdom, households generate 1.05 million tons of plastic packaging waste per year, with 0.63 million ton generated from commercial and industrial sources. In total, 1.68 million tons of plastic packaging have the potential to be fully recycled (Appleton et al., 2005). Polyethylene, polypropylene, polystyrene, poly(vinyl chloride) (PVC), and their blends are the most common plastics that have to be recycled.

The current methods for treating plastic waste aim to either recover energy through incineration and recover chemical value or reuse the plastic. Microwave-induced pyrolysis processes are relatively new and were initially developed by Tech-En Ltd. in Hainault, UK (Lulow-Palafox et al., 2001a). This process involves the mixing of plastic-containing wastes, which are known to have very high transparencies to microwaves, with a highly microwave-absorbent material such as particulate carbon (Lulow-Palafox et al., 2001b). The carbon reaches temperatures around 1,000 °C within a few minutes in the microwave field, and energy is transferred to the shredded plastic by conduction, providing the efficient energy transfer associated with microwave heating processes.

The plastic content of the laminates is transformed into a wide variety of organic compounds that can be used in other chemical processes. Of the total plastic content, about 80% is transformed into oils and waxes and the other 20% into noncondensable gaseous compounds (Lulow-Palafox et al., 2001a). The gaseous fraction can be used in a self-sustaining process by burning the gas and using it as a fuel or energy source.

Poly(vinyl chloride)

Plastic wastes have high heat potentials, but PVC has been considered unfavorable for thermal recycling because of hydrogen chloride release. PVC requires dehydrochlorination in order to be treated by pyrolysis. Because PVC absorbs microwave power more efficiently than other commercially used thermoplastic waste, the application of microwave irradiation for dehydrochlorination of PVC was examined (Moriwaki et al., 2006a). The objective was to elucidate the temperature dependency of microwave power absorption on PVC by observing the temperature profile during irradiation. The scheme of the experimental apparatus is shown in Figure 7.1.

For the waste treatment of PVC by microwave irradiation, the temperature dependency of dielectric loss would be a major factor in the dehydrochlorination process. The measurement shows (Fig. 7.2) that the dielectric loss is increased in proportion with a temperature rise in the first temperature domain; after that, this parameter started to rise quickly at the temperature exceeding 120 °C.

The abrupt rise in dielectric loss and in temperature by microwave irradiation is called thermal runaway. The starting temperature is designated as thermal runaway temperature (TRT). It has been estimated that the TRT is predominantly controlled by dielectric loss of the irradiated material and by microwave power. The relation between incident microwave power and TRT for PVC is shown in Figure 7.3.

The TRT value is the effect of thermal balance of absorbing microwave power and heat removal of the material through the reactor. Based on these results, the peak temperature seems to represent the end point of the dehydrochlorination reaction. The TRT is affected not only by the dependency "dielectric loss versus temperature rise" but also by other irradiation conditions such preheating temperature and microwave power, which are presented in Figure 7.4.

More than the 90% of dehydrochlorination yield is achieved by microwave irradiation onto PVC polymer, indicating that preheating the material resulted in shorter irradiation time to reach such yield.

In a similar paper, microwave irradiation was applied for the removal of chlorine from PVC (Ito et al., 2006). The thermal exposure of pure PVC was studied at high temperatures, and the relationship between the thermal exposure time and residual weight was systematically observed. The decomposition of PVC by 2.45-GHz microwave irradiation was performed using a

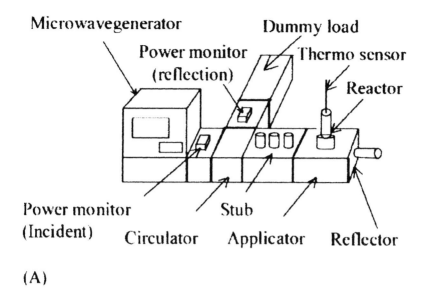

(A)

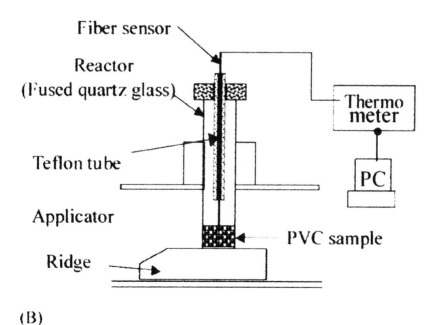

(B)

Fig. 7.1.   Schematic drawing of experimental apparatus. (A) Experimental set-up.
(D) Schematic of temperature measurement. Reprinted from: Moriwaki, S.,
Machina, M., Matsumoto, H., Kuga, M., Ogura, T. 2006a. J. Anal. Appl. Pyrolysis
76:238, with permission.

231

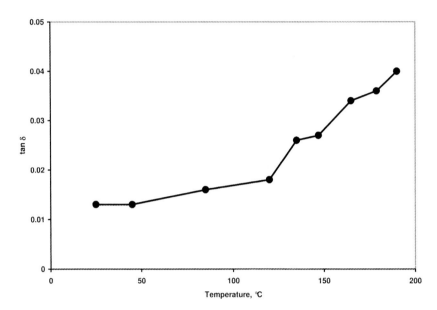

Fig. 7.2. Loss factor of commercial PVC resin plate at 2.45 GHz. Reprinted from Moriwaki, S., Machina, M., Matsumoto, H., Kuga, M., Ogura, T. 2006a. J. Anal. Appl. Pyrolysis 76:238, with permission.

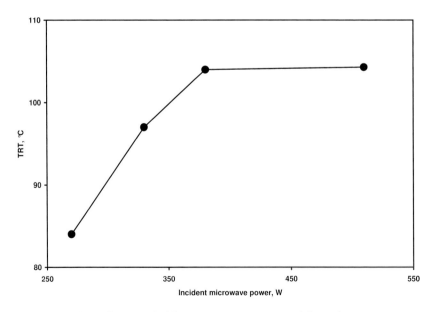

Fig. 7.3. Relation between incident microwave power and thermal runaway temperature (TRT) on PVC polymer. Reprinted from Moriwaki, S., Machina, M., Matsumoto, H., Kuga, M., Ogura, T. 2006a. J. Anal. Appl. Pyrolysis 76:238, with permission.

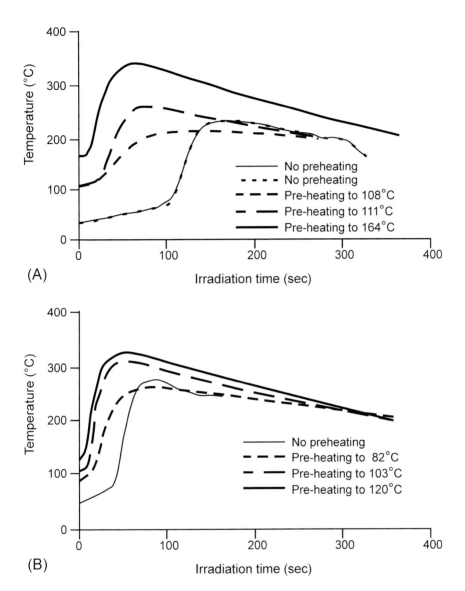

Fig. 7.4. Temperature profile as a function of time at low and high microwave power for different microwave power and different preliminary temperature conditions. (**A**) At low microwave power approximately 360-W irradiation. (**B**) At high microwave power approximately 600-W irradiation. Reprinted from Moriwaki, S., Machina, M., Matsumoto, H., Kuga, M., Ogura, T. 2006a. J. Anal. Appl. Pyrolysis 76:238, with permission.

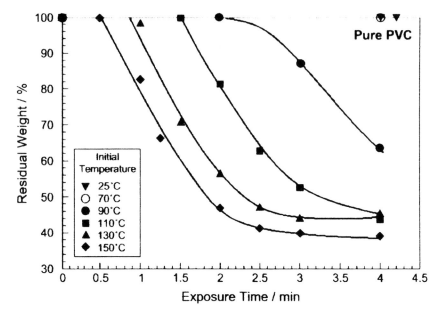

Fig. 7.5.   Weight decrease curve of pure PVC on microwave irradiation after pre-heating at various temperatures. Reprinted from Ito, M., Ushida, K., Nakao, N., Kikuchi, N., Nozaki, R., Asai, K., Washio, M. 2006. Polym. Degrad. Stab. 91:1694, with permission.

domestic microwave oven applying a moderate electrical field intensity. The microwave sensitivity was enhanced above T_g or softening tempera-ture using low frequency. The dechlorination efficiency was markedly en-hanced by preheating of the sample. The microwave absorption efficiency and the weight loss of PVC sample as a consequence depend on the pre-heating temperature before irradiation. The decomposition curves are shown in Figure 7.5.

Apparently, preheating reduced the period before the weight loss started. Figure 7.6 presents the decrease in the weight of WC-PVC (EW-602C; ap-proximately 50 wt% of pure PVC compounded with esters of phthalic acid and calcium carbonate) used as a jacket material for electric cables.

In this case, the decomposition starts after a shorter period, compared to the samples of pure PVC preheated at the same temperature.

In order to ensure the high dehydrochlorination rate of pure PVC and re-productive of the examination result, activated carbon and Zn-Mn ferrite powder, which shows high loss factor for microwave energy absorption, were used as microwave absorbents (Morirwaki et al., 2006b). The use of such absorbents affected the PVC decomposition process, and the effects of decomposition ratio versus microwave energy are shown in Figure 7.7.

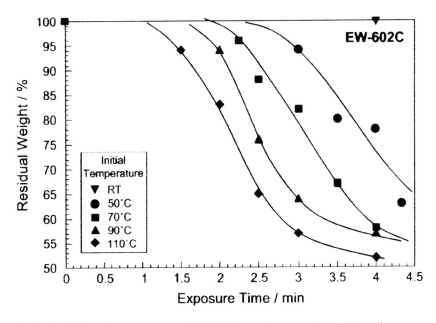

Fig. 7.6. Weight decrease curve of WC-PVC on microwave irradiation after pre-heating at various temperatures. Reprinted from Ito, M., Ushida, K., Nakao, N., Kikuchi, N., Nozaki, R., Asai, K., Washio, M. 2006. Polym. Degrad. Stab. 91:1694, with permission.

It was shown that the activated carbon is superior to the ferrite of microwave energy.

Dehydrochlorination results of commercial PVC resins (listed in Table 7.1) are presented in Figure 7.8.

The dehydrochlorination ratio varies by PVC grade and irradiated microwave energy. The higher content of PVC in plastics resulted in higher yield of material decomposition.

Polystyrene

The kinetics of the oxidative degradation under microwave irradiation of polystyrene in dichlorobenzene in the presence of benzoyl peroxide were studied (Sivalingam et al., 2003). It was shown that microwave irradiation also considerably enhances oxidative degradation of polymers that are carried out in a solution. The microwave-assisted oxidative degradation of polystyrene was investigated at different heating times ranging from 20 to 45 seconds, while several cycles of heating were applied. Figure 7.9 shows

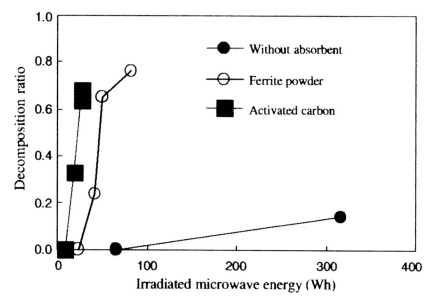

Fig. 7.7. The effect of microwave energy of decomposition ratio on PVC polymer for different microwave energy absorbents. Reprinted from Moriwaki, S., Machina, M., Matsumoto, H., Otsubo, Y., Aikawa, M., Ogura, T. 2006b. Applied Thermal Engineering 26:745 with permission.

Table 7.1   The specification of PVC resin sample.

Substance	Polymerization degree	PVC content determined by analysis, %	Additives
Binder holder	800	82.5	Sn-stabilizer
Flooring material	500–800	6.0	Phthalic plasticizer
Hose	2200	48.2	Ba/Zn-stabilizer
Pipe cramp	700	63.7	Sn-stabilizer

Reprinted from Moriwaki, S., Machina, M., Matsumoto, H., Otsubo, Y., Aikawa, M., Ogura, T. 2006b. Applied Thermal Engineering 26:745, with permission.

the rapid reduction in number average molecular weight of polystyrene with peroxide concentration of 25 kg/m³ in the initial 6 minutes of the effective microwave exposure. Under conventional thermal heating conditions, appreciable degradation of polystyrene was not observed.

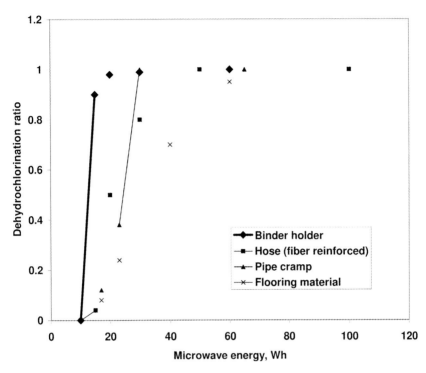

Fig. 7.8. Relation between dehydrochlorination ratio and irradiated microwave energy on PVC resin wastes. Reprinted from Moriwaki, S., Machina, M., Matsumoto, H., Otsubo, Y., Aikawa, M., Ogura, T. 2006b. Applied Thermal Engineering 26:745, with permission.

Poly(ethylene oxide)

Microwave-assisted oxidative degradation was also used in the case of poly(ethylene oxide) (PEO). The process was carried out in a domestic microwave oven with a maximum power of 700 W (Vijayalakshmi et al., 2005). The concentration of polymer solution in water was 2 g/L, and the reaction was carried out in a 50-mL glass beaker. The effects of potassium persulfate concentration, reaction temperature, and different microwave heating cycle times on the degradation rate of PEO were studied. In order to ensure uniform heating and to avoid temperature gradient, the samples were rotated on a turntable. When the polymer solution was exposed to microwave irradiation in the presence of persulfate, rapid degradation occurred, whereas no degradation was observed without the oxidative agents. The degradation of PEO in the microwave oven was compared with degradation under thermal heating conditions. Thus, the solution of PEO with a

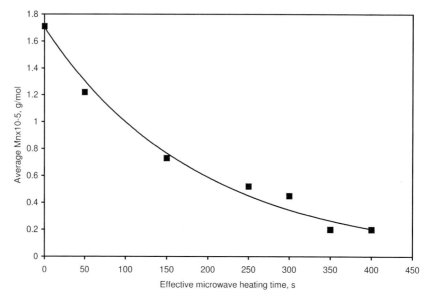

Fig. 7.9. Variation of the number molecular weight of polystyrene with peroxide concentration of 25 kg/m³ and 25-second heating cycle in the microwave. Reprinted from Sivalingam, G., Agarwal, N., Madras, G. 2003. AIChE Journal 49:1821, with permission.

persulfate concentration of 1 g/L was heated at the highest temperature reached in the microwave for a period of 2 minutes. It was concluded that at the comparable temperature range, a microwave effect existed that enhanced the rate of degradation because no significant degradation of PEO was observed under conventional conditions. Variation of Mn_o/Mn versus time for 1 g/L persulfate concentration and 10-, 15-, and 20-second heating cycles is shown in Figure 7.10.

The rate of degradation increased with the temperature, persulfate concentration, and heating cycle time. Moreover, it was stated that lower activation energies, higher rates, and reduced reaction times are attractive features that microwave-assisted degradation offers (Vijayalakshmi et al., 2005).

Polyamide

The depolymerization of polyamide-6 (PA-6) was performed using microwave irradiation with phosphoric acid as a catalyst (Klun et al., 2000). Besides its catalytic activity, phosphoric acid has a very high dipole moment, which makes it an excellent microwave absorbent. Reaction mixtures

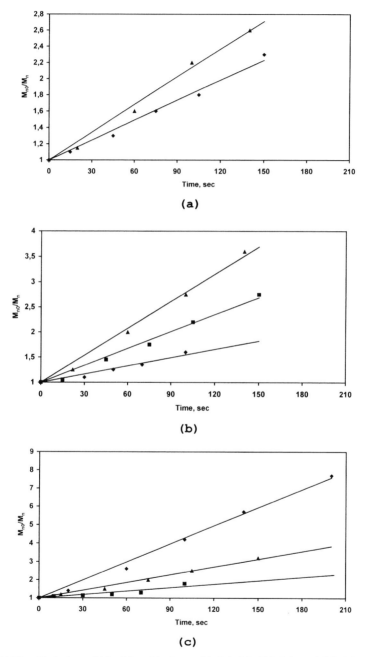

Fig. 7.10. Variation of M_{n0}/M_n with time with (**a**) 0.2, (**b**) 0.4, and (**c**) 1.0 g/L persulfate: (●) 10-, (■) 15-, and (▲) 20-second heating cycle. Reprinted from Vijayalakshmi, S., Chakraborty, J., Madras, G. 2005. J. Appl. Polym. Sci. 96:2090, with permission.

consisted of 10 g of PA-6, 10 g of water, and 1 to 10 g of concentrated H_3PO_4. Commercial PA-6 was grounded to 0.5- to 1-mm particles and used without further preparation. Microwave irradiation of 200 W was applied for 23 minutes to the reaction mixture in a sealed reaction vessel in a multimode microwave reactor. After 15 minutes of irradiation, PA-6 was completely solubilized, and the solubilization efficiency increased linearly up to 6 g when phosphoric acid was added. However, the acidolysis was most effective with 7 g of the acid. The resulted product mixture consisted of more than 90% ε-aminocaproic acid (ACA) and its linear oligomers together with the minor part consisting of cyclic products. With longer irradiation time or more acid, there was a shift toward higher concentration of ACA and its dimmer (Fig. 7.11) as well as a decrease in the concentration of higher oligomers.

Polyesters and Polyurethanes

Poly(ethylene terephthalate)

The safe disposal of poly(ethylene terephthalate) (PET) and polyurethane (PUR) scraps demands new solutions for saving energy and costs. Cross-

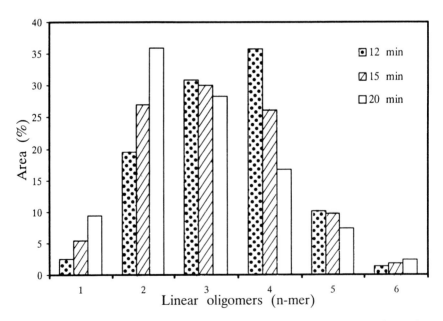

Fig. 7.11.   Relative content of linear oligomers in depolymerization products obtained with irradiation times of 12, 15, and 20 minutes and 9 g of acid addition. Reprinted from Klun, U., Krzan A. 2000. Polymer 41:436, with permission.

linked plastic scraps, including most of PURs, need special technologies for the recovery of polymer components (Randall et al., 2002; Scheirs, 2001).

Solvolysis by glycols and alcohols is an established method for the chemical recycling of PET. The use of microwave irradiation as an energy source in PET solvolysis reactions and the conditions that govern its effectiveness were presented (Krzan, 1998). PET was obtained from washed, used beverage bottles that were chopped into flakes varying from 1 to 15 mm. Solvolysis reagents were reagent grade: methanol, propylene glycol (PG), and polyethylene glycol (PEG). Zinc acetate was used as a catalyst. Degradation experiments were performed in a multimode microwave reactor. A typical sample contained 0.75 g of PET flakes, 0.1 g of zinc acetate, and 10.0 g of the solvolysis reagent mixed in the PTFE vessel. Finally, the characterization of the resulting solution was performed by gel permeation chromatography. It was found that complete PET degradation-solubilization was achieved in the shortest time (i.e., 4 minutes) in methanol, while more than 8 minutes was required in PEG. PET consumption versus microwave irradiation time in a closed system with different alcohols is shown in Figure 7.12.

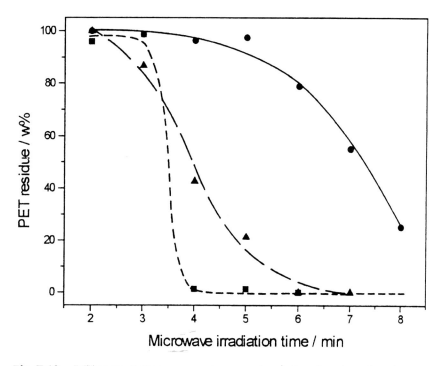

Fig. 7.12. PET consumption versus microwave irradiation time in a closed system with (■) methanol, (▲) propylene glycol, and (•) polyethylene glycol 400 as solvolysis reagents. Applied microwave power was 500 W during all experiments. Reprinted from Krzan, A. 1998. J. Appl. Polym. Sci. 69:1115, with permission.

The investigation showed that the use of microwave radiation as the energy source in PET solvolysis resulted in the short reaction times (a few minutes) needed for complete PET degradation, compared with conventional heating methods (a few hours) (Krzan, 1998).

In the next study, the glycolysis reactions of PET in diethylene glycol (DEG) and PG were monitored in terms of temperature and pressure (Krzan, 1999). A number of basic inorganic compounds ($NaHCO_3$, K_2CO_3, CaO, KH_2PO_4, $NaOCH_3$) were examined as potential catalysts. The reactions were carried out in a multimode microwave reactor equipped with temperature and pressure sensors inserted directly into the reaction vessel. For both DEG and PG, temperature and pressure in the reaction vessel were measured during the exposure of reaction mixtures to microwave irradiation (500 W) for 5 minutes. Maximum temperature of approximately 240 and 200 °C and maximum pressure of about 1.0 and 1.6×10^5 Pa were detected for DEG and PG, respectively. The results were in agreement with an average molecular weight of 556 g/mol reported for polyols obtained by depolymerization of PET with DEG under conventional heating. The method was compatible with the use of a wide range of glycol reagents as well as a variety of basic catalysts. The application of microwave irradiation was not limited to using only zinc acetate or other acetates as catalysts. With many other compounds, comparable effectiveness can be achieved. The main advantage of microwave application was a short reaction time for complete alcoholysis and glycolysis of PET is needed (Krzan, 1999).

PET hydrolytic depolymerization experiments were also carried out under microwave irradiation (Liu et al., 2005). The reactions were run in multimode microwave reactors equipped with temperature and pressure sensors that could be inserted directly into the 100-mL reaction vessel. A typical sample contained 2 g of PET flakes and 20 mL of distilled water, which were put in the reaction vessel without stirring. Experimental conditions of the hydrolytic depolymerization of PET are shown in Table 7.2.

The influence of the applied irradiation power on the depolymerization degree was negligible. However, the microwave irradiation power affected the time needed to reach the set reaction temperature as shown in Figure 7.13.

It was found that the complete depolymerization of PET to terephthalic acid and ethylene glycol was achieved at a pressure of 20 bar and within 120 minutes in pure water without any catalyst.

Recently, the glycolysis of PET under microwave irradiation was investigated in higher scale: 200 g of PET flakes and weight ratio of DEG to PET in the range of 0.5 to 2 (Bogdal et al., 2005). The reason to keep such a ratio of DEG to PET was to achieve a complete conversion of DEG and to obtain such recyclates that can be applied for further processing (e.g., the synthesis of PUR foams or unsaturated polyester resins) without additional

Table 7.2. Experimental conditions of the hydrolytic depolymerization of PET.

Water/ PET	P, bar	t, min	Power, W	Degree of PET depolymerization, %	TPA yield, %
5	15	120	600	75.54	75.82
8	15	120	600	80.24	79.15
10	15	120	600	84.39	84.09
15	15	120	600	72.35	73.14
20	15	120	600	45.18	44.78
10	10	120	600	21.56	21.82
10	13	120	600	34.52	34.59
10	18	120	600	95.42	95.40
10	20	120	600	100	100
10	15	50	600	2.82	2.80
10	15	60	600	10.47	11.04
10	15	70	600	30.20	31.15
10	15	80	600	45.74	44.87
10	15	90	600	75.31	75.69
10	15	100	600	82.50	83.27
10	15	120	600	84.39	84.09
10	20	50	600	9.64	10.15
10	20	60	600	69.36	70.61
10	20	70	600	93.85	94.15
10	20	90	600	99.03	98.94
10	20	120	600	100	100
10	15	120	200	84.94	84.70
10	15	120	400	84.33	84.00
10	15	120	600	84.39	84.09
10	15	120	800	84.24	83.80

P, reaction pressure; t, reaction time.
Reprinted from Liu, L., Zhang, D., An, L., Zhang, H., Tian, Y. 2005. J. Appl. Polym. Sci. 95:719, with permission.

steps of purification and/or concentration of the recyclates. The reaction was carried out in a 500-mL reaction flask, in which 200 g of PET flakes were mixed with DEG and a catalyst. The flask was placed in a multimode microwave reactor in which an upright condenser was mounted and filled with an inert gas. Then, the flask was irradiated up to 230 to 240 °C, and the reaction mixture was stirred by means of a magnetic stirrer.

It was found that a reaction time of 25 to 30 minutes was long enough to dissolve PET flakes in the reaction mixture. Further increase in the reaction time up to 60 minutes did not influence the recyclate hydroxyl numbers, which remained in the range of 530 to 580 mg KOH/g. The analysis

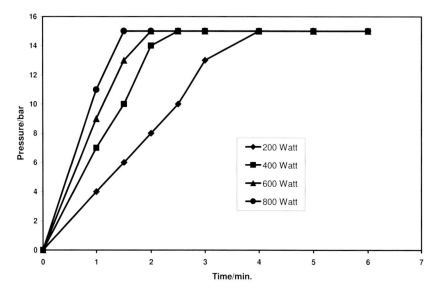

Fig. 7.13. Pressure versus time of different microwave irradiation powers at the initial heating stage. Reprinted from Liu, L., Zhang, D., An, L., Zhang, H., Tian, Y. 2005. J. Appl. Polym. Sci. 95:719, with permission.

of molecular weight distributions of the recyclates revealed that they can be characterized into four fractions with the number average molecular weight of approximately 1,110, 835, 575, and 106 g/mol; however, it was hard to completely separate the fractions 1,110 and 835 g/mol, and they were counted together.

The fraction distribution was dependent on the ratio of DEG to PET and amount of catalyst but did not change substantially when the reaction time was increased over 30 minutes. The increase in the weight ratio of DEG to PET resulted in the increase in the heaviest fractions (i.e., 1,110 and 835 g/mol), while the fraction 335 g/mol was reduced substantially (Fig. 7.14). However, the increase of DEG in the reaction mixture caused the increase of the fraction 106 g/mol (i.e., the increase of DEG in the recyclates).

The catalyst was applied in the range of 0.05 to 1 wt% of PET taken to glycolysis. However, as can be expected, the higher concentrations of the catalyst in the reaction mixtures resulted in the decrease in the participation of heavier fractions, but the fraction distribution was not influenced with the rise of the reaction time from 30 to 60 minutes (Fig. 7.15).

Dependent on the expected viscosity of the recyclates, which depends on the DEG-to-PET ratio, the optimal conditions for the PET glycolysis were to maintain the reaction mixture for 30 minutes at 230 to 240 °C of microwave irradiation in the presence of 0.5 wt% of the catalyst.

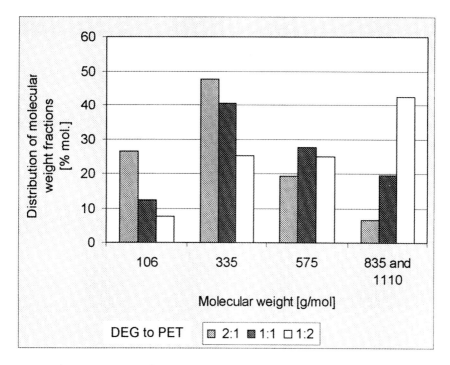

Fig. 7.14. Distribution of molecular weight fractions for PET glycolysis versus the ratio of DEG to PET.

Polyurethanes (PUR)

Most of PUR scraps are in the form of porous materials, which can constitute up to 15% to 20% by weight of total cellular PUR production. Today, the substantial portion of PUR scraps are still disposed of in landfills, and this leads to a loss of material value of the PUR plastic. Therefore, alternative economical means are sought, and among these, physical and chemical recycling represent promising means. Physical recycling of PUR is already established using rebound and regrind technologies. Both these options have specific processing limitations and, therefore, the potential market is limited (Prociak et al., 2002; Scheirs, 2001).

Crosslinked plastic scraps, including most of the PURs, need special technologies for the recovery of polymer components. In the processes of advanced chemical recycling, methods such as hydrolysis, aminolysis, and glycolysis of PUR wastes can be used to recover the polyols (Modesti, 1996). At present, glycolysis appears to be the most favorable method for chemical recycling of PUR waste on an industrial scale. In this way, PUR

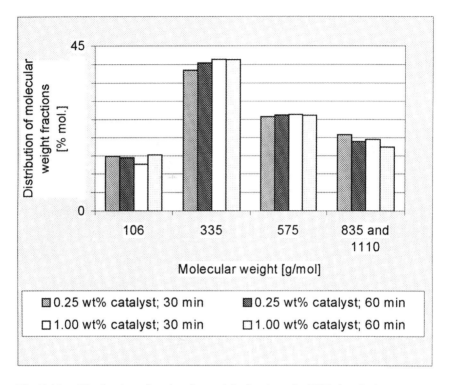

Fig. 7.15.   Distribution of molecular weight fractions for PET glycolysis versus catalyst concentration.

scraps are liquefied and products are obtained that contain terminal hydroxyl groups (Fig. 7.16) capable of forming new PUR materials.

In general, glycolysis involves the treatment of manufacturing scraps with excess of a glycol at temperatures of 150 to 250 °C. Catalysts are usually used in these processes (Frisch, 1998; Prociak et al., 2003). The possibilities of applying the recovered polyols to formulate modern ecological systems for manufacturing heat-insulating PUR foams are discussed in a number of works (Modesti, 1996; Prociak, 2004; Prociak et al., 2003).

The method of limited depolymerization of postmanufacture PUR scraps allows possible reusing of glycolysis products in new formulations. The application of these glycolysates can be different, but use of them for preparing insulating foams is particularly interesting due to their long-term applications (Prociak et al. 2002, 2003). In conventional process, PUR scraps of microporous elastomers or rigid foams were ground before being fed into a hot glycol solution. The use of low-molecular-weight glycols (ethylene glycol, DEG) allowed a relatively high ratio of PUR to glycol, which is economically favorable. The glycolysis process under conventional heating

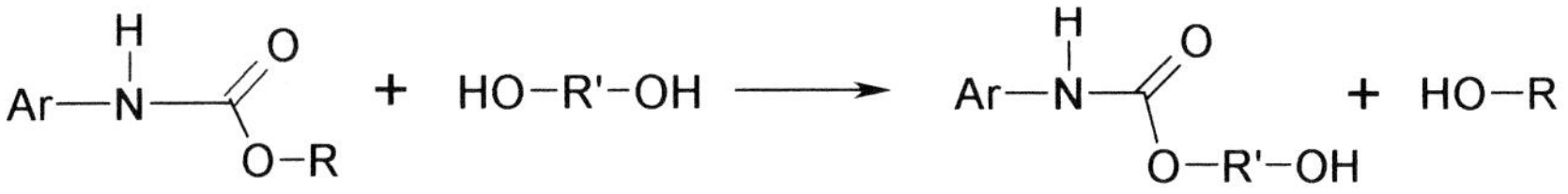

Fig. 7.16. Glycolysis reaction of polyurethane waste.

was carried out within 2 to 6 hours with a weight ratio of PUR to glycol from 1:1 to 10:1 with or without a catalyst.

Thus, the investigation of glycolysis reactions of PUR systems under microwave irradiations was carried out by applying microporous elastomer and rigid foams of PUR as starting materials (Bogdal et al., 2003).

At the beginning of the investigation, the glycolysis process was carried out with a molar ratio of PUR scraps to monoethylene glycol (MEG) of 1:1 and resulted in relatively high hydroxyl numbers of final products, which in turn created some difficulties during their reapplication to PUR systems.

Therefore, in the next stages of the work, it was decided to use twofold excess of PUR scraps so that final products were characterized by hydroxyl numbers of approximately 400 mg KOH/g. Such polyols can be applied to the composition of PUR systems for the production of rigid foams. Eventually, the application of microwave irradiation to the PUR scraps recycling allowed glycolysates to be obtained in a fourfold shorter time (approximately 30 minutes), but the properties of glycolysates were the same as under conventional conditions (Bogdal et al., 2004). Mass average molecular weights of glycolysates versus reaction time for the PUR glycolysis under microwave conditions are shown in Figure 7.17.

It can be seen that under microwave irradiation, the strongest reduction in molecular weights of the reaction mixture as a result of the highest rate of PUR decomposition was during the first 20 minutes of the process. After 30 minutes it was slower to reach desired values. However, for the process with a catalyst, mass average molecular weights decreased faster to reach the desired value after 15 minutes; the reaction time of 30 minutes was needed to get rid of small pieces of solid material (PUR) present in reaction mixtures (Fig. 7.18). Similar observations were made for the reaction under conventional conditions, where the minimum time required for the desired homogenization of reaction mixture was 2 hours.

Finally, the glycolysates that were prepared under conventional and microwave conditions were added to the PUR formulations to manufacture rigid foams for insulating applications.

In a typical procedure under microwave conditions, the PUR scrapes were mixed together with DEG in a weight ratio of foam or elastomer scraps to DEG of 2:1, placed in a flask, and irradiated under an upright con-

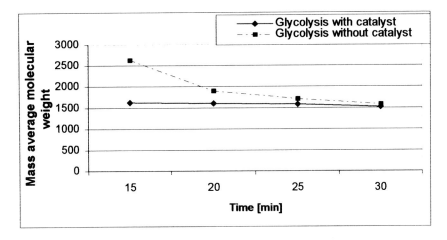

Fig. 7.17.   Mass average molecular weights of glycolysates versus reaction time for the PUR glycolysis under microwave conditions. Reprinted from: Bogdal, D., Prociak, A., Pielichowski, J. 2004. Global Symposium on Recycling, Waste Treatment and Clean Technology, REWAS 2004, Madrid, vol. III, 2807, with permission.

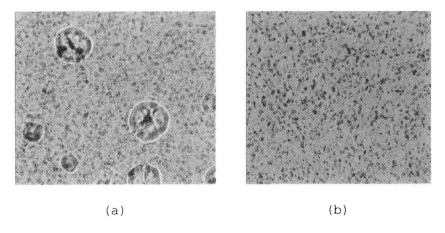

(a)   (b)

Fig. 7.18.   Microscopic image of glycolysis products obtained under (**a**) thermal and (**b**) microwave conditions.

denser for 30 minutes with or without a catalyst. Because various types of PUR foams are manufactured on the base of different raw materials, each type of scrap demands an elaboration of different conditions for a glycolysis process. The postconsumer scraps of PUR elastomer for footwear and rigid foams were used, and after glycolysis, they were added instead of neat

Table 7.3. The characteristic of the liquid product of the polyurethane elastomer degradation.

Process conditions	Conventional process	Process in microwave reactor
Temperature, °C	199–201	190–200
Power, W	250	300
Reaction time, h	2	0.5
Product properties		
Colour	Cream	Cream
Homogeneity	Good	Very good
Density, g/cm^3, 25 °C	1, 1	1, 1
Viscosity, mPa × s, 25 °C	2200	1560
Hydroxyl number, mg KOH/g	440	280

Reprinted from Bogdal, D., Prociak, A., Pielichowski, J. 2004. Global Symposium on Recycling, Waste Treatment and Clean Technology, REWAS 2004, Madrid, vol. III, 2807, with permission.

polyols in PUR systems for the preparation of rigid foams that exhibited satisfactory thermal insulating properties (Bogdal et al., 2004).

The glycolysis product obtained with DEG was used to substitute, in the amount up to 50 wt%, the petrochemical polyol mixture in new PUR compositions for foaming heat-insulating materials. Advantageous effects of the additions of 15 to 30 wt% of the glycolysate on the foam properties such as compressive strength and thermal conductivity were found. The addition of glycolysis product in the amount exceeding 40 wt% led to the formation of foams that were unfortunately useless for further practical application (Prociak et al., 2005).

It was found that due to lower viscosity of glycolysates obtained under microwave conditions, blowing processes of PUR systems were characterized by a similar set of parameters as for foams in which only neat polyols were used (Bogdal et al., 2004), while with glycolysates obtained under conventional conditions, the mixing process during formulation of the PUR system was slightly more difficult. The characteristic of the liquid products of the foam's limited degradation under microwave and conventional conditions is presented in Table 7.3.

The glycolysis processes were also carried out using the different ratio of waste to DEG in the range of 2:1 to 10:1 (Prociak et al., 2006). The temperature changes in the reaction mixture during the glycolysis process of 150 g of components (PUR waste and DEG) are shown in Figure 7.19.

The reaction mixture temperature of approximately 200 °C in a multi-

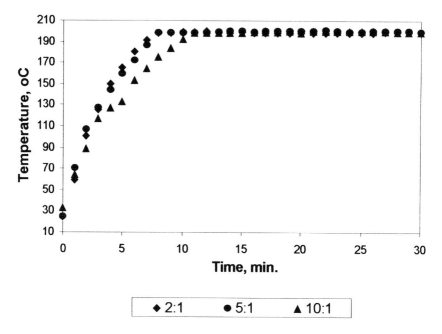

Fig. 7.19.   The temperature changes of reaction mixture during the glycolysis process of 150-g components (PUR waste and DEG).

mode microwave reactor (maximum power of 600 W) was acquired after 8 to 10 minutes, and the distinct difference in the time to reach this temperature was not observed for various ratios of PUR waste to glycol.

The scaling-up procedure of reaction mixture weights was also investigated between 150 g and 450 g. The dependence of the reaction mixture temperature on microwave irradiation time is shown in Figure 7.20 for the same mixing ratio (i.e., 10:1) of PUR waste to DEG. It was clearly visible that in the case of the 450-g sample, the temperature of approximately 200 °C was reached after a threefold longer heating period in comparison to the sample with 150-g weight.

When the reaction mixture of approximately 200 °C was reached, a similar power was needed to support the temperature on the same level as is shown in Figure 7.21.

To sum up, the developed processes of glycolysis under microwave conditions and novel PUR systems have many ecological and economical advantages:

- Total recirculation of all scraps
- Short time of glycolysis process without a catalyst

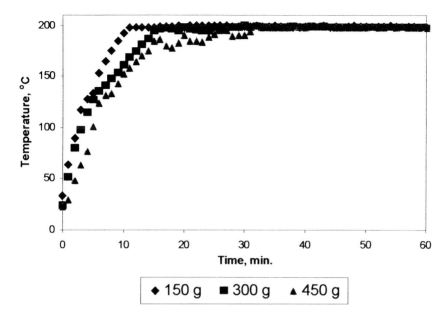

Fig. 7.20. The dependence of reaction mixture temperature on microwave irradiation time.

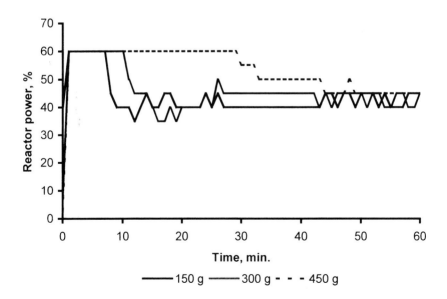

Fig. 7.21. Power need to support the temperature of approximately 200 °C of the reaction mixture (PUR waste and DEG).

- Heating reaction mixture in whole volume by means of microwave
- Saving energy and cost
- Replacement of origin polyols with glycolysis products
- Saving valuable raw materials
- Using new generation blowing agents
- Low density of foams
- Low coefficient of thermal conductivity

Rubber

Considerable research has been carried out on the potential recycling of tires. Scrap tires are amenable to thermal treatment, which apart from incineration, includes gasification and pyrolysis. Tires are a composite mixture of materials that upon vulcanization and coupling with a wire support form the tire (Sharma et al., 1998). The mixture comprises the following:

- Elastomer (natural or synthetic rubber)
- Reinforcing agents (carbon black)
- Plasticizers (hydrocarbon oils)
- Vulcanizing agents (sulfur and sulfur compounds)
- Accelerating agents (to facilitate sulfur action)
- Protective agents (antioxidizing, antioxonant, stabilizer)

Conventional pyrolysis is used to convert scrap tires into value-added products, such as olefins, chemicals, and surface-activated carbons. The technique permits volumetric reduction, material recovery, and can render harmless many organic substances.

Waste tire processing can provide the following (Jones et al., 2002):

- Reuse of the rubber material (to produce products similar to the initial ones)
- Recovery of materials and energy

Due to the vulcanization process in the manufacture of rubber, used tires are not directly reusable. It is necessary to treat the rubber with a devulcanization process for regeneration and reuse as virgin materials (Sharma et al., 1998). Devulcanization breaks the sulfur carbon bonding that makes rubber stable. Alternatively, tires can be milled, either mechanically or cryogenically, to give an equally manageable granular material or powder. The milling process tends to be cheaper than the devulcanization one. However, with a microwave heating process, the correct amount of energy can be used to cleave carbon-carbon bonds without depolymerization occurring (Adhikari et al., 2000). The presence of polar compounds is a prerequisite for the microwave devulcanization to be effective. An example is sulfur-vulcanized elastomers that contain such polar groups and, therefore, are suitable for this method. A pyrolysis plant that employs microwave energy to break down the tires has been developed (Robinson, 1989) by BRC Environmental Limited. They proposed a revised plant to apply microwave pyrolysis to the recovering of components in scrap tires. The plant would process some 50,000 tons of scrap tires per annum to produce carbon, steel cord, and hydrocarbon oils that could be sold and a mixture of hydrogen and methane gases that could be used to fuel the process. Note the similarity to conventional pyrolysis. Apart from process water, the plant would not produce any emissions. Earlier attempts at the new technique failed due to insufficient temperatures that resulted in incomplete pyrolysis of the forming char. BRC's method uses microwave energy to heat tires more uniformly and quickly to 2,000 °C. A carbon recovery of 36% is claimed, with the main product (33%) being the highly valued activated carbon the char product fraction. Other products include steel cord and hydrocarbon oils, while the remaining component, a mixture of hydrogen and methane gases, helps to power the heating process. As the process is carried out in a closed system (inert atmosphere of nitrogen), the production of dioxins, soot, and ash is prevented.

BRC Environmental stopped their work on microwave pyrolysis of scrap tires in the mid-1990s. The literature suggests that further developments in this field have only been done recently. The UK-based company Amat has developed a new microwave system (Lee, 2003). The system can be used for reprocessing scrap tires, plastics, and petroleum. Details of the system are sparse within the literature; however, it was suggested that the technique can provide successful treatment, although there are reports that, due to the more conventional techniques currently being favored, it is unlikely that the system will be employed in the United Kingdom in the near future. Any plant intending to treat tires to obtain valuable products must overcome economic barriers. Consequently, a process is required that offers low-temperature pyrolysis but at an acceptable process time. Both cannot be accomplished using conventional techniques. Although microwave pyrolysis takes comparably less time, process temperatures are higher. Investigations should look toward reducing treatment temperatures to yield high-molecular-weight olefins (Appleton et al., 2005).

Carbon Fiber Composites with Epoxy Resins

A common problem with existing plastic recycling processes is that the plastics are often composed of more than one polymer or there can be

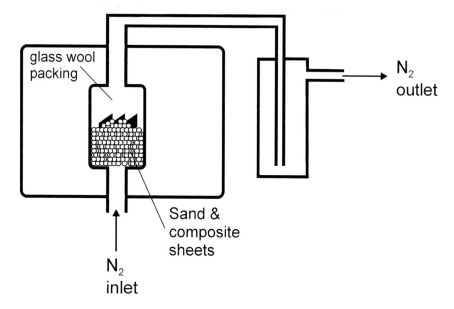

Fig. 7.22.   A schematic of the experimental rig. Reprinted from Lester, E., Kingman, S., Wong, K., Rudd, C., Pickering, S., Hilal, N. 2004. Mater. Res. Bull. 39:1549, with permission.

fibers added to form a composite to give added strength. Current and impending legislation provides additional motivation for energy solutions; rising landfill taxes and the Europe Union end-of-life vehicle directive mean that saving energy solutions are required urgently.

Carbon fiber composites with an epoxy resin matrix were subjected to microwave heating experiments in order to volatilize the polymer content and to produce clean fibers for potential reuse in high-grade applications (Lester et al. 2004). The composites were processed at 3 kW for 8 seconds in a multimode microwave applicator (Fig. 7.22).

Four 3-g sheets of composite were suspended in a bed of quartz sand, because sand is essentially transparent to microwaves. Glass wool was used to prevent solids leaving the microwave cavity through the glassware. A steady stream of nitrogen gas was used to create an inert atmosphere and to prevent combustion of the fibers during heating. The tensile strength and modulus of virgin treated in microwave and in fluidized bed (at 450 °C) fibers were compared in Table 7.4. It may also be worth noting that the modulus of the microwave-heated samples was slightly reduced compared with both the virgin and fluidized bed samples.

It appears technically feasible to recycle the polymer as well as the carbon fiber.

Table 7.4. Mechanical properties of fibers.

Property	Virgin fibers		Treated fibers	
	Manufacture data[a]	Experimental data	Fluidized bed	Microwaved
Tensile strength, GPa	4.5	4.09	3.05	3.26
Tensile modulus, GPa	234	242	243	210

[a] Grafil 34-700 standard modulus carbon fiber data sheet.
Reprinted from Lester, E., Kingman, S., Wong, K., Rudd, C., Pickering, S., Hilal, N. 2004. Mater. Res. Bull. 39:1549 with permission.

References

Adhikari, B., De, D., Maiti, S. 2000. Prog. Polym. Sci. 25:909.

Appleton, T., Colder, R., Kingman, S., Lowndes, I., Read, A. 2005. Applied Energy 81:85.

Bogdal, D., Penczek, P., Pielichowski, J., Prociak, A. 2003. Adv. Polym. Sci. 163:193.

Bogdal, D., Prociak, A., Pielichowski, J., Balazinski, K. 2005. Waste Recycling. IGSMiE PAN, Krakow, Poland.

Bogdal, D., Prociak, A., Pielichowski, J. 2004. Global Symposium on Recycling, Waste Treatment and Clean Technology, REWAS 2004, Madrid, vol. III, 2807.

Frisch, K. 1998. Polimery 10:579.

Ito, M., Ushida, K., Nakao, N., Kikuchi, N., Nozaki, R., Asai, K., Washio, M. 2006. Polym. Degrad. Stab. 91:1694.

Jones, D., Lelyveld, T., Mavrofidis, S., Kingman, S., Miles, N. 2002. Resources, Conservation and Recycling 34:75.

Klun, U., Krzan A. 2000. Polymer 41:4361.

Krzan, A. 1998. J. Appl. Polym. Sci. 69:1115.

Krzan, A., 1999. Polym. Advan. Technol. 10:603.

Lee, A. 2003. The Engineer 291:12.

Lester, E., Kingman, S., Wong, K., Rudd, C., Pickering, S., Hilal, N. 2004. Mater. Res. Bull. 39:1549.

Liu, L., Zhang, D., An, L., Zhang, H., Tian, Y. 2005. J. Appl. Polym. Sci. 95:719.

Lulow-Palafox, C., Chase, H. 2001a. Chem. Eng. 717:28.

Lulow-Palafox, C., Chase, H. 2001b. Ind. Eng. Chem. Res. 40:4749.

Modesti, M. 1996. Advances in Urethane Science and Technology 13:237.

Moriwaki, S., Machina, M., Matsumoto, H., Kuga, M., Ogura, T. 2006(a). J. Anal. Appl. Pyrolysis 76:238.

Moriwaki, S., Machina, M., Matsumoto, H., Otsubo, Y., Aikawa, M., Ogura, T. 2006(b). Applied Thermal Engineering 26:745.

Prociak, A. 2004. First International Seminar on Modern Polymeric Materials for Environmental Applications, Krakow, Seminar Proceedings, 121.

Prociak, A., Bogdal, D., Pielichowski, J., Janikowska, A. 2006. In preparation.

Prociak, A., Pielichowski J. 2005. Polimery 50:682.

Prociak, A., Pielichowski, J. 2002. III International Conference "Plastics Recycling", Jesenik, Czech Republic:207.

Prociak, A., Pielichowski, J. 2003. Przem. Chem. 82/8-9: 951.

Prociak, A., Pielichowski, J. 2004. Global Symposium on Recycling, Waste Treatment and Clean Technology, REWAS 2004, Madrid, vol.III:2351.

Randall, D., Lee, S. (editors) 2002. The Polyurethanes Book, J. Wiley & Sons Ltd.

Robinson, S. 1989. European Rubber Journal 171:32.

Scheirs, J. 2001. Polymer Recycling, J. Wiley & Sons Ltd.

Sharma, V., Mincarini, M., Fortuna, F., Cognini, F., Cornacchia, G. 1998. Energy Convers. Manage. 39:511.

Sivalingam, G., Agarwal, N., Madras, G. 2003. AIChE Journal 49:1821.

Vijayalakshmi, S., Chakraborty, J., Madras, G. 2005. J. Appl. Polym. Sci. 96:2090.

8

COMMERCIALIZATION AND SCALING-UP

From the chapters already presented, it can be seen that there are a lot of factors that influence polymer synthesis as well their processing under microwave conditions. Not all materials are good microwave absorbers and, therefore, suitable for microwave applications, which in turn results in a unique treatment of every process that has to be matched for it. Thus, a real cross-disciplinary approach has to be considered to fully understand all the limitations and advantages of microwave processing. Improper application of microwave irradiation will usually lead to disappointments, while proper understanding and use of microwave power can bring even greater benefits than can be expected.

For example, microwave heating was rapidly adopted by the food industry in the early 1950s. The first means of conveying materials through a microwave cavity was patented in 1952 (Spencer, 1952). However, it awaited the development, by Cryodry in 1962, of successful means for a choking conveyor tunnel opening before food-processing applications became practical. For nearly 40 years following the invention of microwave oven in 1945, industrialists and microwave equipment manufacturers predicted a rosy future for microwave processing, while very few expected that microwave ovens would became popular home appliances; reality was quite different.

In 1984, the whole of the food industry applied 19 MW of microwave heating units installed worldwide (Decareau, 1986). Compared with the 5.1 MW of microwave heating units in use in 1978 (Bengtsson et al., 1980), the annual growth in the years 1978 to 1984 was on the order of 2.5 MW per year, or an average increase of 20%. In addition, the number of industrial installations built by users for their own processes were not included in this statistic. In absolute terms, the 19 MW of industrial microwave power was equivalent to 32,000 domestic microwave ovens of 600 W, which is derisory by comparison to the approximately 7 million units in domestic service in

Table 8.1.   Comparison of consumer and industrial microwave heating markets in the United States (all numbers are approximate).

	Consumer ovens	Industrial system
Numbers of units installed	180 million	<500
Total sales value ($US)	27 billion[a]	500 million[b]
Installed megawatts (output)	108.000[c]	<100

[a] $150 per oven.
[b] $3000–$5000 per kW output.
[c] 600 W per oven.
From Schiffmann, R.F. 2001. State of the art of microwave applications in the food industry in the USA. 8th International Conference on Microwave and High Frequency Heating. Bayreuth. Germany.

1984 in the United States, where they showed the highest growth rate in the electric domestic appliances sector between 1976 and 1986 (Thuery, 1992). In fact, industrial microwave food processing has never achieved the success of domestic microwave ovens, in which speed and convenience of reheating were powerful driving forces for consumers. However, speed and convenience have little meaning in an industry where the financial bottom line and return of investment are the most powerful motivators for adopting new processes (Table 8.1) (Schiffmann, 2001).

During the past four decades, it has become apparent that there is an overriding need for close cooperation between the user and the manufacturer of microwave equipment. Often, the manufacturer of microwave equipment has little knowledge of the industrial process being broached, while on the other hand, the user does not understand the basic principles of microwave heating and apparatus. It is only the convergence of these two interests that can result in a successful application. A prerequisite for this is clearly that each must be the master of his discipline. The past decades or so have brought more reliable microwave equipment. However, equally important is the appreciation by the user of the energy costs of his existing process (Metaxas et al., 1983).

On the other hand, the use of microwave heating in an industrial scope usually has quite different requirements compared to the application in a research environment. When looking into microwave research programs, conferences, or papers, the nearly total absence of industrial microwave heating applications or, at least, industrial-scale research is notable. When looking into these sources, one could get the impression that microwave heating applications are researched but rarely applied in production. The greatest differences are the motivations of the project and magnitude of the equipment.

While the motivation to use microwaves of a research application may be manifold, ranging from fundamental research to finding new applications, the industrial intention for using microwaves or other new technology is usually only a single one—profit. Therefore, the industrial project is limited to giving answers to two questions (Moller et al., 2005). Is it possible to use microwaves for the intended process? If yes, is it more profitable to use microwaves for the process than the existing technology?

In fact, microwaves do not provide a universal solution to all problems but should be considered whenever all other processes fail to solve an industrial problem, in which case the advantages become unique and offer considerable savings compared to other existing processes. Since magnetron efficiencies are between 50% and 72% for 2.45 GHz compared to 80% and 87% for high-power 922 or 915 MHz, a necessary but not sufficient condition then is that the product to be treated must be of high quality, so that the cost of microwave treatment can be justified. The study of efficiency should also include a detailed analysis of the conventional process, of which the user often has a very vague idea (Thuery, 1992).

The main advantages and disadvantages of microwave processing in comparison with conventional processes are listed by AMT (Advanced Material Technology). Among the main advantages of microwave processes are counted the following (AMT, 2006):

- Volumetric heating that gives rise to a *very rapid energy transfer* into the material being heated. In conventional heating, heat flow is initiated on the materials surface, and the rate of heat flow into the center is dependent on the materials thermal properties and the temperature differential. A conventional oven is required to be heated to temperatures much higher than is required by the material itself since their asymptotical rise in workload temperature toward the required level.
- *Energy savings*, along with the fact that in a properly designed applicator the majority of the available energy is dissipated in the workload, lower temperatures associated with the cavity surroundings mean that radiation, conduction, and convection heat losses are reduced. This can represent energy savings of up to 70%.
- *Instantaneous control of power*, giving better control of process parameters and rapid start-up and shut down.
- *Reduced equipment size*; the rapid dissipation of energy, mainly into the workload, and the high-energy densities capable in small volumes allow equipment to be up to 20% the physical size of conventional systems.
- *Selective heating*; the material ability to be heated by electromagnetic energy is dependent on its dielectric properties, which this means that

in a mixture containing a number of differing constituents, the heating of each will vary. This can have profound positive consequences on energy use, bulk reaction temperatures, moisture removal, and process simplification.

- *Clean energy transfer;* the nature of microwaves means that energy transfer to a material is usually via some form of polarization effect within the material itself. This direct transfer of energy eliminates many of the problems associated with organic fuel usage for the end user.
- *Chemical reactions driven;* as has been shown in this book a number of chemical reactions can be accelerated using microwaves. Solvent-free reactions are gaining popularity in many labs, thus reducing problems associated with waste disposal of solvents and other hazardous chemicals.

Among the most serious disadvantages of microwave processes are counted the following (AMT, 2006):

- *Field complexity;* it can be very difficult to accurately predict the exact nature of electromagnetic field interaction with materials. This difficulty is especially true when using multimode cavities with their many different field patterns possible. However, this unpredictability is often easily overcome and understood by experimentation, and many successful industrial applications have been implemented without the need to know precise field behavior.
- *Temperature uniformity* can be influenced by many factors but, broadly speaking, uniformity is a function of either material characteristics or boundary conditions.
- *Temperature measurement;* the spot nature of most temperature measurement techniques and the inherent nonuniform heating of electromagnetic heating mean that a spot measurement can be misleading.
- *Initial capital cost;* compared to conventional heating techniques, electromagnetic heating in industrial situations usually requires a higher level of initial capital outlay.
- *Largest magnetron;* the largest single microwave source suitable for industrial applications is approximately 100 kW. Applications requiring large amounts of energy would need multiple sources.
- *R & D cost;* all new applications usually require development of the microwave applicator and understanding of the fundamentals associated with the new process. This means that immediate implementation is not usually possible.

Thus, the process characteristics that have been suggested potentially attractive for microwave processing are as follows (NAP, 1994):

Table 8.2. Capital cost of industrial microwave equipment.

Component	Typical cost
Complete system	$1.000–5000/kW
Generator	<50% of system cost
Applicator	>50% of system cost
Power transmission	($1.000–3.000), ≤5% of system cost
Instrumentation	($1.000–3.000), ≤5% of system cost
External materials handling	($1.000–3.000), ≤5% of system cost
Installation and start-up	5–15% of system cost

Source: Sheppard, 1988; reprinted from NAP. 1994. Microwave Processing of Materials. The National Academy Press with permission.

- The size or thickness of the material should be large.
- The cost of the material should be high.
- Improvements in properties obtainable from microwave processing are significant.
- Plant space is limited.
- Electricity is cheap.
- Minimizing handling is advantageous.

Other characteristics may include:

- Heat from the combustion of coal, oil, or natural gas is not practical (i.e., electricity is the only power source).
- Maintaining a very clean, controlled processing environment is important.

Economic feasibility is a function of local variations in energy costs, environmental laws, and labor costs balanced with the properties of finished materials or parts, improvements in yield or productivity, and the markets for the products. In this chapter, only estimations of capital cost and operating costs of industrial microwave equipment are given (Sheppard, 1988). Due to the differences in the configurations and processing approaches between microwave and conventional systems, it is very difficult to perform a general comparison of capital costs in a meaningful way. However, microwave processing equipment is almost always more expensive than conventional systems (Table 8.2) (NAP, 1994).

Operating costs include the cost of energy, both absolute and the real cost per part based on the coupling efficiency and the size and number of parts, and the costs of maintenance, repair, and replacement. Table 8.3 gives some estimates of these costs. In summary, the economic benefits of microwave processing are difficult to define in a general way. The decision

Table 8.3. Operating costs of microwave equipment.

Component	Typical cost
Magnetron replacement	1–12 c/kW × h
Electric energy	5–12 c/kW × h
Plug-to-product efficiency	
915 MHz	70–75%
2.450 MHz	50–65%
Routine maintenance	5–10%

Source: Sheppard, 1988; reprinted from: NAP. 1994. Microwave Processing of Materials. The National Academy Press with permission.

to use microwave processing for any application has to be based on an analysis of the specific process. Important factors include the location of the processing facility, the product requirements, possible property improvements, alternative sources of energy, availability of capital, and the balance between energy costs, labor costs, capital costs, and the value added to the product (NAP, 1994).

From this very simplified overview, it becomes clear why the results of a research project are not much use for an industrial project of even the same topic. Although much of the process generated in understanding a research project is not required for the average industrial project, it should nevertheless be possible to come to at least the same result as the industrial project. The other big difference between a research project and an industrial one is the size and power of the microwave test unit used (Moller, 2005). Therefore, it is worth mentioning what equipment is available industrially and for laboratory uses.

As stated in Chapter 2, the microwave-assisted vulcanization of rubber compounds is the most important application of microwave heating to polymeric materials in terms of number of installed plants. At the moment, microwave equipment manufacturers offer single-mode as well as multi-mode microwave ovens, which are designed for automobile type profile and big industrial type profiles, respectively. Present implementations are characterized by 3- to 10-kW applied power, shifting velocity in the range of 0.75 to 45 m/min. Treated profile of approximately 50 × 50 mm passes the 3-m-long microwave chamber followed by a 7-m-long space of hot air followed by a cooling section of approximately 4 m. Result processing capacity is about 500 kg/h, enabling processing of profiles with metal frames, skeletons, and reinforcements. However, profile section, properties of given types of rubber, and required production capacity affect the speed of profile advance; control of the process, and especially the extent of in-

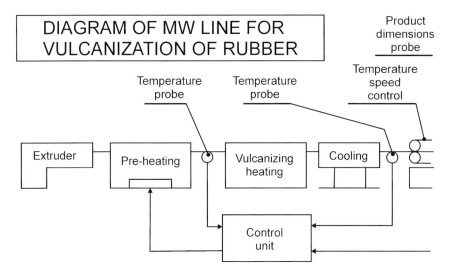

DIAGRAM OF MW LINE FOR VULCANIZATION OF RUBBER

Product dimensions probe

Temperature probe

Temperature probe

Temperature speed control

Extruder

Pre-heating

Vulcanizing heating

Cooling

Control unit

Fig. 8.1 Diagram of microwave line for vulcanization of rubber. Reprinted from Romill. 2006. www.romill.cz, with permission.

stalled microwave output heat, is typically 1 kW per approximately 30 kg of the product per hour (Fig. 8.1) (Romill, 2006).

The temperature in the extruder rises to 80 to 90 °C, and microwave preheating, placed before the input of the vulcanization tunnel, is set to approximately 130 °C. After entering the tunnel, the material is quickly heated to the vulcanization temperature, e.g., 180 °C, depending on the mix composition. After passing through the microwave tunnel, the material is maintained for 60 to 90 seconds at the required temperature by the hot-air system and then cooled. Microwave vulcanization under atmospheric pressure allows universal design and processing of various profiles without fundamental modification of lines, of which construction is mostly similar (Romill, 2006).

Microwave irradiation is also used for rubber recycling, which requires disrupting the bridges on the vulcanization level and sustaining the bridges on the polymerization level. The major difficulty is the fact that devulcanization and depolymerization temperatures are quite close to each other. This is due to the similarity of atomic structure to be broken: S-S and C-S bonds are disrupted during devulcanization, while C-C bounds are disrupted in depolymerization. That is why very high levels of uniformity and homogeneity of the process are required in the whole volume. For instance, such a devulcanization line had been functioning at Goodyear, in which 70-kW equipment processed the 280-kg/h output. Devulcanization is followed by extrusion and 15% to 20% virgin rubber addition. Finally, it is applied on inner tire surfaces. Compared with conventional devulcanization, the process better preserves the mechanical properties. The costs at 0.33 USD/kg are proved to be lower, as well as easy integration into the existing continuous technology and lower environmental impact (Thuery, 1992).

For processing of polymeric materials and composites, a number of industrial microwave equipment manufacturers offer equipment for the production of continuous cast-resin components, in which the microwave unit (3.6 kW or 7.2 kW) processes high-viscosity resins systems with flow rates up to 5.0 kg/min. The control system provides easy integration into other and/or existing systems. Several furnaces can be switched in cascades to achieve an increase in temperature difference between feeding flow and drain flow and/or an increase in the flow quantity of the medium to be treated. The microwave flow heater is available, which can also be applied in other fields, such as the food, plastic, and chemical industries. There is a microwave continuous heating chamber for glass fiber cables, glass fiber–reinforced plastics, and reinforced optical fibers (0.8, 1.6, 2.4 kW) (Linn, 2006).

There also are continuous microwave belt furnaces, in which the microwave generators are arranged in a spiral around the longitudinal axis of the cylindrical chamber to achieve a more uniform energy distribution. The conveyor belt passes over floor plates, which are fitted with secondary radiators to provide a higher microwave concentration. The furnaces can be operated with a microwave power output of up to 200 kW. The power control system allows each magnetron to be switched on and off if necessary, which offers the advantage that power consumption is reduced compared to continuous power control. Size of the channel opening is adapted to a process (Fig. 8.2) (Linn, 2006).

Recently, an industrial microwave system, HEPHAISTOS-CA2 (High Electromagnetic Power Heating Autoclaveless Injected STructures Oven System), for curing of carbon fiber–reinforced plastics was developed (Feher et al., 2006). The system is especially optimized for processes like injection molding or curing of matrix and modular system technology in connection with "autoclaves" fabrication processes.

As previously stated, the second segment in the microwave chemistry market is for instruments for laboratory use in chemical synthesis and analysis that target customers at academic working groups and laboratories in the chemical, pharmaceutical, and biochemical industries. The intention of all manufacturers as well as users is to extend the dimensions of product capacity. Thus, most research in the field of chemical synthesis has been recently focused on scaling up of chemical reactions using microwave irradiation. Manufacturers of microwave instruments have devised means of

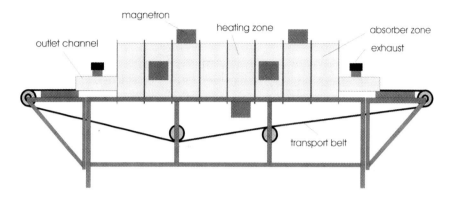

Fig. 8.2   Continuous microwave belt furnace. Reprinted from Linn High Therm.
2006. www.linn.de, with permission.

scaling up the yield of both single-mode and multimode microwave reactors. Based on the patent publications granted to date in this field, it is evident that research is still in a nascent stage (Taylor et al., 2006). Detailed descriptions of such instruments can be found in the recent review papers and chapters (Bogdal, 2005; Kappe et al., 2005; Ondruschka et al., 2006). Some information is also given in Chapter 1.

Recently, a continuous microwave system that consists of four microwave cavities with a rotating quartz tube was tested and developed (Bogdal, 2006). All the microwave cavities are equipped with a continuous power regulation and temperature control, while each magnetron can be separately switched on and off if necessary, which offers the advantage of reduced power consumption compared to continuous power control. The speed of the quartz tube rotation is also continuously adjustable. The system was designed for the investigation of the recycling of polymeric materials but recently has also been applied for chemical synthesis (Fig. 8.3).

More recently, Novartis designed and built a microwave work station that is equipped with four single-mode microwave reactors, capping and decapping stations, robotic arm, transport and rack storage system, pipetting robot with DMSO and acetone stations, and drying and gassing stations. The microwave instruments are directly driven by the software via an Ethernet connection, while the scheduler maximizes the throughput of the system by allowing parallel multitasking (Chamoin, 2006).

Over the next 5 years, the market for microwave instruments for chemical analysis and synthesis is predicted to grow at a CAGR (i.e., the year-over year growth rate of an investment over a specified period of time) of 9.6% per annum, to reach 145.8 million USD by 2008. The chemical synthesis market will give a major thrust to the microwave chemistry market,

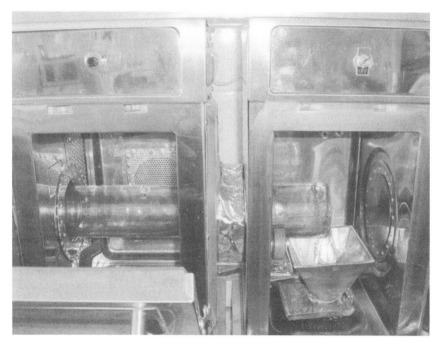

Fig. 8.3   Continuous microwave reactor with rotating quartz tube (from Politechnika Krakowska).

which is expected to grow at a CAGR of 20% per year, to 67.2 million USD by 2008. As the analytical segment has matured, it is predicted to show a stable CAGR of about 5% per year, reaching 78.6 million USD during the same period. In the next 5 to 6 years, the chemical synthesis segment is expected to overtake the analytical segment in terms of market share. This is primarily due to the increasingly successful implementation of microwave instruments in chemical synthesis. Another contributor will be growing research and intellectual property activity in the segment, such as research on the scaling up of reactions (Taylor et al., 2005).

However, as was stated in the Evalueserve report, information on market statistics pertaining to microwave equipment is not available freely, so it was not possible to make a comprehensive analysis of the microwave chemistry market. This also limited research to freely available online sources. As a result, certain key information, deemed necessary for a comprehensive analysis of the microwave chemistry market, could not be obtained through primary research. This was because some respondents did not want to reveal this information due to confidentiality issues (Taylor et al., 2005)

The overview of microwave equipment manufacturers is presented in Table 8.4 (Kubel, 2005). Many companies that specialize in microwave

Table 8.4. Microwave equipment manufacturers.

Microwave furnace manufacturers	Specially	Web address
Advanced Manufacturing Technology	Industrial microwave heating consultants	www.amtmicrowave.com
Anton-Paar	Microwave digestions systems and reactors	www.anton-paar.com
Biotage	Microwave system organic synthesis, biochemistry	www.biotage.com
CEM Corporation	Microwave reactors and digestions systems	www.cem.com
CM Furnaces	Microwave + electric hybrid furnaces	www.cmfurnaces.com
Cober-Muegge	Microwave vulcanization, equipment	www.cobermuegge.com
Communications and Power Industries	Microwave Heating systems and equipment	www.cpii.com
C-Tech Innovations/ Ceralink	Design and build microwave hybrid furnaces and other mw furnaces	www.ceralink.com
Dennis Tool	Continuous Microwave Sintering Furnaces	
Ertec	Microwave reactors and systems for digestion and analysis	www.ertec.pl
Ferrite Components, Inc.	Microwave tempering and cooking systems	www.ferriteinc.com
Gerling Applied Engineering	Microwave systems and equipment	www.2450mhz.com
Harper International	Microwave Rotary Calciner	www.harperintl.com
Harrop Industries	Microwave gas hybrid furnaces	www.harropusa.com
Industrial Microwave Systems, Inc. (US)	Microwave systems continuous planar, cylindrical	www.industrialmicrowave.com
Industrial Microwave Systems, Inc. (UK)	Continuous Microwave Furnaces and equipment	www.industrial-microwave-systems.com
Linn High Therm GmbH	Dual Frequency microwave furnaces	www.linn.de

(continued)

Table 8.4.   Microwave equipment manufacturers. (*continued*)

Microwave furnace manufacturers	Specially	Web address
Manitou Systems Inc.	Microwave and RF plasma systems	www.manitousys.com
Microdry Corp.	Industrial microwave systems	www.microdry.com
Milestone S.R.L.	Microwave systems digestion and analysis	www.milestonesci.com
Mino Yogyo Co. Ltd.	Microwave hybrid furnaces	www.mini-ceramic.co.jp
O-I-Corporation	Microwave digestion systems	www.oico.com
Panasonic	Microwave pottery kiln	
Plazmatronika	Microwave reactors and systems for digestion and analysis	www.plazmatronika.pl
PSC	Radio frequency heating systems and microwave equipment	www.pscrfheat.com
Puschner-Microwave Power Systems	Microwave furnace, dryers	www.pueschner.com
Radatherm Pty Ltd	Microwave sintering furnaces	www.radatherm.com.au
Research Microwave Systems, LLC	Laboratory microwave systems, accessories	www.thermwave.com
Romill spol. s r.o.	Microwave systems and equipments	www.romill.cz
Sairem	Radio frequency heating systems and microwave equipment	
Takasago Industry Co	Microwave batch and microwave elevator kiln	www.takasago-inc.co.jp
Thermex-Thermatron, Inc.	High power, high frequency MW and RF heaters	www.thermex-thermatron.com

Reprinted from Kubel, E. 2005. Industrial Heating. 43, with permission.

power will build microwave equipment to customer specification, and some will assist with design and processing issues. However, industry must have some understanding of how microwaves interact and heat materials and what the limitations are before its potential can adapted.

In the production circumstances, the design of microwave equipment must be stimulated by the end users, and equipment manufacturers need to be ready to understand and respond to this new market. On the other hand, microwave equipment manufacturers can encourage this new microwave market by making standard research and production systems available (Kubel, 2005).

To sum up, fundamental differences in the heat transfer during material processing in thermal and microwave fields is that microwave energy, in contrast to thermal heating, is supplied directly to a large volume, thus avoiding the thermal lags associated with conduction and/or convection. Consequently, temperature gradients and the excessive heat build-up during thermal processing could be reduced by a microwave power control. Thus, a comparison of thermal and microwave processing assumes a new dimension when the temperature distribution inside the sample is considered, and that is where the scientific challenge lies. Regardless of the type of activation (thermal) or kind of microwave effects (nonthermal), microwave energy has its own advantages, which are still waiting to be fully understood and applied to chemical processes.

References

Akyel, C., Bilgen, E. 1989. Energy 14:839.

AMT. 2006. http://www.amtmicrowave.com/pros_cons.html.

Bengtsson, N., Ohlsson, T. 1980. Application of Microwave and High-Frequency Heating in Food Processing. In: Food Process Engineering, Linko, P., Malkki, Y., Olkko, J., Larinkari, J. (eds.). Applied Science Publishers. London.

Bogdal, D. 2005. Microwave-Assisted Organic Synthesis, One Hundred Reaction Procedures. Elsevier. Amsterdam.

Bogdal, D. 2006. Unpublished data.

Chamoin, S. 2006. High-Throughput Microwave Synthesis at Novartis. Advances in Microwave-Assisted Organic Synthesis. MAOS-Conference and Exhibition. Budapest.

Decareau, R. 1986. Food. Technol. 40:99.

Feher, L., Thumm, M., Drechsler, K. 2006. Adv. Eng. Mater. 8:26.

Kappe, O., Stadler, A. 2005. Microwave in Organic and Medical Chemistry. Wiley-VCH. Weinheim.

Krieger, B. 1992. Polym. Mater. Sci. Eng. 66:339.

Krieger, B. 1994. Mater. Res. Soc. Symp. Proc. 347:57.

Kubel, E. 2005. Industrial Heating. 43.

Linn High Therm. 2006. www.linn.de.

Metaxas, A.C., Meredith, R.J. 1983. Industrial Microwave Heating. Peter Perigrinus. London.

Moller, M., Linn, H. 2005. Industrial Microwave Heating. An Underestimated Technology. 10th International Conference on Microwave and High Frequency Heating. Modena. Italy

National Academic Press. 1994. Microwave Processing of Materials. The National Academy Press.

Ondruschka, B., Bonrath, W., Stuerga, D. 2006. Development and Design of Laboratory and Pilot Scale Reactors for Microwave-assisted Chemistry. In: Microwaves in Organic Chemistry. Loupy, A. (ed.). Weinheim. Wiley-VCH.

Romill. 2006. www.romill.cz.

Schiffmann, R.F. 2001. State of the Art of Microwave Applications in the Food Industry in the USA. 8th International Conference on Microwave and High Frequency Heating. Bayreuth, Germany.

Sheppard, L.M. Ceramic Bull. 67:1556.

Spencer, P.L. 1952. US Patent 2 605 383.

Taylor, M., Atri, B.S., Minhas, S. 2005. Development in Microwave Chemistry. Evalueserve. RSC.

Thuery, J. 1992. Microwaves: Industrial, Scientific, and Medical Applications. Artech House. Boston/London.

INDEX

271